MATLAB for Beginners

A Gentle Approach

Revised Edition

Peter I. Kattan

Petra Books

www.PetraBooks.com

Peter I. Kattan, PhD

Correspondence about this book may be sent to the author at one of the following two email addresses:

pkattan@petrabooks.com

info@petrabooks.com

MATLAB for Beginners: A Gentle Approach,
Revised Edition, written by Peter I. Kattan.
ISBN: 979-8-8690-5591-0

In Loving Memory of My Father

MATLAB for Beginners
A Gentle Approach
Revised Edition

Preface

This book is written for people who wish to learn MATLAB[1] for the first time. The book is really designed for beginners and students. In addition, the book is suitable for students and researchers in various disciplines ranging from engineers and scientists to biologists and environmental scientists. The book is intended to be used as a first course in MATLAB in these areas. Students at both the undergraduate level and graduate level may benefit from this book. The material presented has been simplified to such a degree such that the book may also be used for students and teachers in high schools – at least for performing simple arithmetic and algebraic manipulations. One of the objectives of writing this book is to introduce MATLAB and its powerful and simple computational abilities to students in high schools.

The material presented increases gradually in difficulty from the first few chapters on simple arithmetic operations to the last two chapters on solving equations and an introduction to calculus. In particular, the material presented in this book culminates in Chapters 6 and 7 on vectors and matrices, respectively. There is no discussion of strings, character variables or logical operators in this book. The emphasis is on computational and symbolic aspects. In addition, the MATLAB Symbolic Math Toolbox[2] is emphasized in this book. A section is added at the end of each chapter (except chapters 1 and 9) dealing with various aspects of algebraic and symbolic computations with the Symbolic Math Toolbox. In fact, the final chapter on calculus assumes entirely the use of this toolbox.

---

[1] MATLAB is a registered trademark of the MathWorks, Inc.
[2] The MATLAB Symbolic Math Toolbox is a registered trademark of the MathWorks, Inc.

Chapter 1 provides an overview of MATLAB and may be skipped upon a first reading of the book. The actual material in sequence starts in Chapter 2 which deals with arithmetic operations. Variables are introduced in Chapter 3 followed by mathematical functions in Chapter 4. The important topic of complex numbers is covered in Chapter 5. This is followed by Chapters 6 and 7 on vectors and matrices, respectively, which provide the main core of the book. In fact, MATLAB stands for MATrix LABoratory – thus a long chapter is devoted to matrices and matrix computations. An introduction to programming using MATLAB is provided in Chapter 8. Plotting two-dimensional and three-dimensional graphs is covered in some detail in Chapter 9. Finally, the last two chapters (10 and 11) present solving equations and an introduction to calculus using MATLAB.

The material presented in this book has been tested with version 7 of MATLAB and should work with any prior or later versions. There are also over 230 exercises at the ends of chapters for students to practice. Detailed solutions to all the exercises are provided in the second half of the book. The material presented is very easy and simple to understand – written in a gentle manner. An extensive references list is also provided at the end of the book with references to books and numerous web links for more information. This book will definitely get you started in MATLAB. The references provided will guide you to other resources where you can get more information.

I would like to thank my family members for their help and continued support without which this book would not have been possible. In this edition, I am providing two email addresses for my readers to contact me - pkattan@petrabooks.com and info@petrabooks.com

October 2023   Peter I. Kattan

Contents

1. Introduction

In this introductory chapter a short MATLAB tutorial is provided. This tutorial describes the basic MATLAB commands needed. More details about these commands will follow in subsequent chapters. Note that there are numerous free MATLAB tutorials on the internet – check references [1-28]. Also, you may want to consult many of the excellent books on the subject – check references [29-47]. This chapter may be skipped on a first reading of the book.

In this tutorial it is assumed that you have started MATLAB on your computer system successfully and that you are now ready to type the commands at the MATLAB prompt (which is denoted by double arrows ">>"). For installing MATLAB on your computer system, check the web links provided at the end of the book.

Entering scalars and simple operations is easy as is shown in the examples below:

```
>> 2*3+5

ans =

    11
```

The order of the operations will be discussed in subsequent chapters.

```
>> cos(60*pi/180)

ans =

    0.5000
```

The argument for the cos command and the value of pi will be discussed in subsequent chapters.

We assign the value of 4 to the variable x as follows:

```
>> x = 4

x =

    4

>> 3/sqrt(2+x)

ans =

   1.2247
```

To suppress the output in MATLAB use a semicolon to end the command line as in the following examples. If the semicolon is not used then the output will be shown by MATLAB:

```
>> y = 30;
>> z = 8;
>> x = 2; y-z;
>> w = 4*y + 3*z

w =

   144
```

MATLAB is case-sensitive, i.e. variables with lowercase letters are different than variables with uppercase letters. Consider the following examples using the variables x and X:

```
>> x = 1

x =
```

```
        1

>> X = 2

X =

        2

>> x

x =

        1

>> X

X =

        2
```

Use the `help` command to obtain help on any particular MATLAB command. The following example demonstrates the use of `help` to obtain help on the `det` command which calculated the determinant of a matrix:

```
>> help det
  DET     Determinant.
     DET(X)  is  the  determinant  of  the  square
matrix X.

     Use  COND  instead  of  DET  to  test  for
matrix singularity.

     See  also  cond.
```

Overloaded functions or methods (ones with the same name in other directories)
        help sym/det.m

    Reference page in Help browser
        doc det

The following examples show how to enter matrices and perform some simple matrix operations:

```
>> x = [1 2 3 ; 4 5 6 ; 7 8 9]

x =

    1    2    3
    4    5    6
    7    8    9

>> y = [1 ; 0 ; -4]

y =

    1
    0
   -4

>> w = x*y

w =

   -11
   -20
   -29
```

Let us now see how to use MATLAB to solve a system of simultaneous algebraic equations. Let us solve the following system of simultaneous algebraic equations:

$$\begin{bmatrix} 3 & 5 & -1 \\ 0 & 4 & 2 \\ -2 & 1 & 5 \end{bmatrix} \begin{Bmatrix} x_1 \\ x_2 \\ x_3 \end{Bmatrix} = \begin{Bmatrix} 2 \\ 1 \\ -4 \end{Bmatrix} \qquad (1.1)$$

We will use Gaussian elimination to solve the above system of equations. This is performed in MATLAB by using the backslash operator "\" as follows:

```
>> A= [3 5 -1 ; 0 4 2 ; -2 1 5]

A =

      3        5       -1
      0        4        2
     -2        1        5

>> b = [2 ; 1 ; -4]

b =

      2
      1
     -4

>> x = A\b

x =

    -1.7692
     1.1154
    -1.7308
```

It is clear that the solution is $x_1 = -1.7692$, $x_2 = 1.1154$, and $x_3 = -1.7308$. Alternatively, one can use the inverse matrix of A to obtain the same solution directly as follows:

```
>> x = inv(A)*b

x =

    -1.7692
     1.1154
    -1.7308
```

It should be noted that using the inverse method usually takes longer than using Gaussian elimination especially for large systems of equations.

Consider now the following 4 x 4 matrix D:

```
>> D = [1 2 3 4 ; 2 4 6 8 ; 3 6 9 12 ; -5 -3
-1 0]

D =

     1     2     3     4
     2     4     6     8
     3     6     9    12
    -5    -3    -1     0
```

We can extract the sub-matrix in rows 2 to 4 and columns 1 to 3 as follows:

```
>> E = D(2:4, 1:3)

E =

     2     4     6
     3     6     9
```

        -5   -3   -1

        We can extract the third column of D as follows:

```
>> F = D(1:4, 3)

F =

        3
        6
        9
       -1
```

        We can extract the second row of D as follows:

```
>> G = D(2, 1:4)

G =

        2    4    6    8
```

        We can also extract the element in row 3 and column 2 as follows:

```
>> H = D(3,2)

H =

        6
```

        We will now show how to produce a two-dimensional plot using MATLAB. In order to plot a graph of the function $y = f(x)$, we use the MATLAB command `plot(x,y)` after we have adequately defined both vectors x and y. The following is a simple example to plot the function $y = x^2 + 3$ for a certain range:

```
>> x = [1 2 3 4 5 6 7 8 9 10 11]

x =

  Columns 1 through 10

      1       2       3       4       5       6       7
  8       9      10

  Column 11

     11

>> y = x.^2 + 3

y =

  Columns 1 through 10

      4       7      12      19      28      39      52
 67      84     103

  Column 11

    124

>> plot(x,y)
```

Figure 1.1 shows the plot obtained by MATLAB. It is usually shown in a separate window. In this figure no titles are given to the x- and y-axes. These titles may be easily added to the figure using the xlabel and ylabel commands. More details about these commands will be presented in the chapter on graphs.

In the calculation of the values of the vector y, notice the dot "." before the exponentiation symbol "^". This dot is used to denote that the operation following it be performed element by element. More details about this issue will be discussed in subsequent chapters.

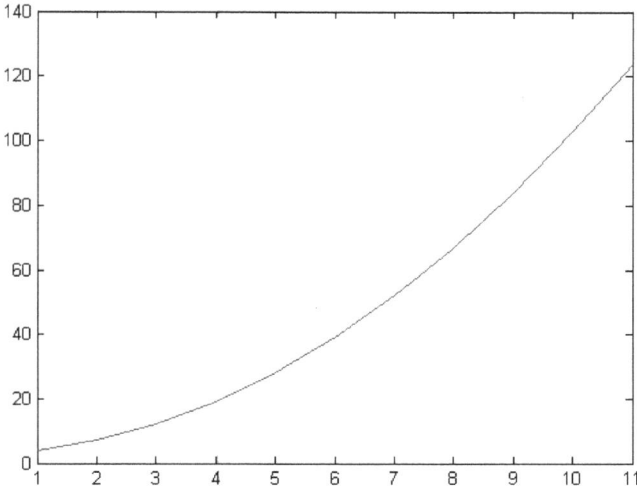

Figure 1.1: Using the MATLAB "plot" command

Finally, we will show how to make magic squares with MATLAB. A magic square is a square grid of numbers where the total of any row, column, or diagonal is the same. Magic squares are produced by MATLAB using the command magic. Here is a simple example to produce a 5 x 5 magic square:

```
>> magic(5)

ans =

    17    24     1     8    15
    23     5     7    14    16
     4     6    13    20    22
    10    12    19    21     3
    11    18    25     2     9
```

It is clear that the total of each row, column or diagonal in the above matrix is 65.

Exercises

Solve all the exercises using MATLAB. All the needed MATLAB commands for these exercises were presented in this chapter.

1. Perform the operation 3*4+6. The order of the operations will be discussed in subsequent chapters.

2. Perform the operation $\cos(5)$. The value of 5 is in radians.

3. Perform the operation $3\sqrt{6+x}$ for $x = 4$.

4. Assign the value of 5.2 to the variable y.

5. Assign the values of 3 and 4 to the variables x and y, respectively, then calculate the value of z where $z = 2x - 7y$.

6. Obtain help on the `inv` command.

7. Generate the following matrix: A

$$A = \begin{bmatrix} 1 & 0 & 2 & -3 \\ 0 & 5 & 2 & 2 \\ 1 & 2 & 3 & 4 \\ -2 & 0 & 1 & 3 \end{bmatrix}$$

8. Generate the following vector b

$$b = \begin{Bmatrix} 1 \\ 2 \\ 3 \\ 4 \end{Bmatrix}$$

9. Evaluate the vector c where $\{c\} = [A]\{b\}$ where A is the matrix given in Exercise 7 above and b is the vector given in Exercise 8 above.

10. Solve the following system of simultaneous algebraic equation using Gaussian elimination.

$$\begin{bmatrix} 5 & 2 \\ 1 & 3 \end{bmatrix} \begin{Bmatrix} x_1 \\ x_2 \end{Bmatrix} = \begin{Bmatrix} 3 \\ -1 \end{Bmatrix}$$

11. Solve the system of simultaneous algebraic equations of Exercise 10 above using matrix inversion.

12. Generate the following matrix X:

$$X = \begin{bmatrix} 1 & 0 & 6 \\ 1 & 2 & 3 \\ 4 & 5 & -2 \end{bmatrix}$$

13. Extract the sub-matrix in rows 2 to 3 and columns 1 to 2 of the matrix X in Exercise 12 above.

14. Extract the second column of the matrix X in Exercise 12 above.

15. Extract the first row of the matrix X in Exercise 12 above.

16. Extract the element in row 1 and column 3 of the matrix X in Exercise 12 above.

17. Generate the row vector x with integer values ranging from 1 to 9.

18. Plot the graph of the function $y = x^3 - 2$ for the range of the values of x in Exercise 17 above.

19. Generate a 4 x 4 magic square. What is the total of each row, column, and diagonal in this matrix.

2. Arithmetic Operations

In this chapter we learn how to perform the simple arithmetic operations of addition, subtraction, multiplication, division, and exponentiation. The first four operations of addition, subtraction, multiplication, and division are performed in MATLAB using the usual symbols +, -, *, and /. The following are some examples:

```
>> 5+6

ans =

    11

>> 12-9

ans =

    3

>> 3*4

ans =

    12

>> 4/2

ans =

    2

>> 4/3

ans =
```

     1.3333

In the next chapter, we will show how to control the number of decimal digits that appear in the answer. The operation of exponentiation is performed using the symbol ^ as shown in the following examples:

```
>> 2^3

ans =

        8

>> 2^-3

ans =

     0.1250

>>
>> -2^3

ans =

      -8

>> -2^4

ans =

     -16

>> (-2)^4

ans =

     16
```

In the above examples, be careful where the minus sign is used. The constants π and ε are obtained in MATLAB as follows:

```
>> pi

ans =

    3.1416

>> eps

ans =

  2.2204e-016
```

In the above examples, the constant π is denoted by the command pi and represents the ratio of the perimeter of a circle to its diameter. The other constant ε is denoted by the command eps and represents the smallest number that MATLAB can handle.

In the above examples, only one arithmetic operation was performed in each line. Alternatively, we can perform multiple arithmetic operations in each line or command. The following are some examples:

```
>> 2+3-7

ans =

    -2

>> 3*4+5

ans =

    17
```

```
>> 2+5/4
```

```
ans =
```

```
    3.2500
```

In these cases, the order of operations is very important because the final computed result depends on this order. Precedence of computation is for the exponentiation operation, followed by multiplication and division, then followed by addition and subtraction. For example, in the examples above, multiplication is performed before addition and division is performed before addition. In case you need to perform addition before multiplication or division, the above two examples may be computed using parentheses as follows:

```
>> 3*(4+5)
```

```
ans =
```

```
    27
```

```
>> (2+5)/4
```

```
ans =
```

```
    1.7500
```

It is clear from the above examples that the operations inside the parentheses take precedence over any other operations. Several multiple operations may also be performed as follows:

```
>> (26+53)-(3^4 +5)*2/3
```

```
ans =
```

```
21.6667
```

One should be careful when using parentheses. A change of the position of the parenthesis in the above example will change the computed results as follows:

```
>> (26+53)-(3^4 +5*2)/3
```

```
ans =
```

```
48.6667
```

Arithmetic Operations with the MATLAB Symbolic Math Toolbox

Symbolic arithmetic operations may also be performed in MATLAB using the MATLAB Symbolic Math Toolbox. For the next examples, this toolbox needs to be installed with MATLAB. In order to perform symbolic operations in MATLAB, we need to define symbolic numbers[3] using the sym command. For example, ½ is defined as a symbolic number as follows:

```
>> sym(1/2)
```

```
ans =
```

```
1/2
```

In the above example, the number ½ is stored in MATLAB without any approximation using decimal digits like 0.5. Once used with the sym command, the computations will be performed algebraically or symbolically. For example, the following addition of two fractions is performed numerically as follows without the sym command:

---

[3] Actually these are symbolic objects in MATLAB but objects are beyond the scope of this book.

```
>> 1/2+3/5

ans =

    1.1000
```

Adding parentheses will not affect the numerical computation in this case:

```
>> (1/2)+(3/5)

ans =

    1.1000
```

However, using the sym command of the MATLAB Symbolic Math Toolbox will result in the computation performed symbolically without the use of decimal digits;

```
>> sym((1/2)+(3/5))

ans =

11/10
```

Notice in the above example how the final answer was cast in the form of a fraction without decimal digits and without calculating a numerical value. In the addition of the two fractions above, MATLAB finds their common denominator and adds them by the usual procedure for rational numbers. Notice also in the above four example outputs that symbolic answers are not indented but numerical answers are indented. In the next chapter, we will study variables and their use in MATLAB.

Exercises

Solve all the exercises using MATLAB. All the needed MATLAB commands for these exercises were presented in this chapter. Note that Exercises 19 and 21 require the use of the MATLAB Symbolic Math Toolbox.

1. Perform the addition operation 7+9.

2. Perform the subtraction operation 16-10.

3. Perform the multiplication operation 2*9.

4. Perform the division operation 12/3.

5. Perform the division operation 12/5.

6. Perform the exponentiation operation 3^5.

7. Perform the exponentiation operation 3*(-5).

8. Perform the exponentiation operation (-3)^5.

9. Perform the exponentiation operation -3^5.

10. Compute the value of $\dfrac{2\pi}{3}$.

11. Obtain the value of the smallest number that can be handled by MATLAB.

12. Perform the multiple operations 5+7-15.

13. Perform the multiple operations (6*7)+4.

14. Perform the multiple operations 6*(7+4).

15. Perform the multiple operations 4.5 + (15/2).

16. Perform the multiple operations (4.5 + 15)/2.

17. Perform the multiple operations (15 − 4 + 12)/5 − 2*(7^4)/100.

18. Perform the multiple operations (15 − 4) + 12/5 − (2*7)^4/100.

19. Define the number 2/3 as a symbolic number.

20. Perform the fraction addition (2/3) + (3/4) numerically.

21. Perform the fraction addition (2/3) + (3/4) symbolically.

3. Variables

In this chapter we introduce variables and how to deal with them in MATLAB. If you do not use a variable, then MATLAB automatically stores the calculated result in the variable ans. The following is an example:

```
>> 3+2

ans =

     5
```

In the above example, the number 5 is stored automatically in the variable ans. This variable is short for "answer". If you wish to store the number 5 in another variable, then you need to specify it as follow:

```
>> x=3+2

x =

     5
```

In the above example, we used the variable x to store the result which is 5. You may define the variables x, y, z as follows without performing any operations:

```
>> x=1

x =

     1

>> y=-3
```

```
y =

    -3

>> z=3.52

z =

    3.5200
```

Note that MATLAB is case-sensitive with respect to variables. For example, the variables x and X are different. The following example illustrates this issue:

```
>> x=5

x =

    5

>> X=3

X =

    3

>> x

x =

    5

>> X

X =

    3
```

We may perform arithmetic operations on variables as shown in the following examples:

```
>> x=4

x =

        4

>> y=2.5

y =

      2.5000

>> z=x+y

z =

      6.5000

>> w=2*x-3*z+y

w =

      -9
```

Up till now, we have used single letters to denote variables. Any string of characters may be used to denote variables also. For example, we next use the variables result and the_answer_is as follows:

```
>> a=3

a =
```

```
        3

>> b=-2

b =

    -2

>> result = a*b

result =

    -6

>> the_answer_is = 2*a/b

the_answer_is =

    -3
```

Since MATLAB is case-sensitive, then the variables `width`, `WIDTH`, and `Width` are all different and store different results. You may add comments to your MATLAB commands and statements by using the symbol %. Comments are not executed and are ignored by MATLAB but are included for the benefit of the reader. For example, here is a comment:

```
>> % This is a comment
```

Comments are displayed in green color on the MATLAB screen. You may also add a comment on the same line as a command. The following is an example:

```
>> x = 6    % the value of x is defined here

x =
```

6

The following are the MATLAB rules for naming variables:

1. MATLAB is case-sensitive with respect to variables.
2. All variables must start with a letter.
3. You may have up to 31 characters in a variable.
4. Punctuation signs are not allowed in a variable.
5. The underscore sign "_" is allowed in a variable.
6. Both letters and digits are allowed in a variable.

For example, we may define variables x1, x2, and x3 as follows:

```
>> x1 = 1

x1 =

     1

>> x2 = -2.7

x2 =

    -2.7000

>> x3 = 2*x1/x2

x3 =

    -0.7407
```

All variable used must be defined. For example, we get an error if we try to use a variable that is not defined. For example, the variable h is not defined as follows:

```
>> 3*h/2
??? Undefined function or variable 'h'.
```

The error is displayed in red color on the MATLAB screen. To find all the variables used in a MATLAB session, use the who command as follows:

```
>> who
```

```
Your variables are:
```

X	result	x1	z
a	the_answer_is	x2	
ans	w	x3	
b	x	y	

The who command does not provide any details about the variables. The following is an example on using variables. If you know the cost and sale price of an item, calculate the profit. Here is the example:

```
>> cost = 100

cost =

    100

>> sale_price = 120

sale_price =

    120

>> profit = sale_price - cost

profit =

    20
```

Numbers in MATLAB are displayed with four decimal digits by default. For example:

```
>> 7/3

ans =

    2.3333
```

However, MATLAB performs the calculations internally using about 16 digits. You may ask MATLAB to display results with 16 digits using the command `format long` as follows:

```
>> format long
>> 7/3

ans =

    2.33333333333333
```

```
>> format short
>> 7/3

ans =

    2.3333
```

The command `format short` tells MATLAB to display four decimals again. Finally, you may ask MATLAB to show all variables used in a session with details regarding size, bytes[4], and class[5] using the `whos` command as follows:

---

[4] The unit "bytes" is a measure of how much storage the variables take in the computer memory or disk. For example, a kilobyte (KB) is one thousand bytes and a megabyte (MB) is one million bytes.

[5] The "class" of a variable indicates the type of the variable. For example, all the variables in this session are of the class "double array". By default, all variables are stored as arrays (i.e. matrices) in MATLAB.

```
>> whos
  Name           Size      Bytes  Class

  X                        1x1      8    double array
  a                        1x1      8    double array
  ans                      1x1      8    double array
  b                        1x1      8    double array
  cost                     1x1      8    double array
  profit                   1x1      8    double array
  result                   1x1      8    double array
  sale_price               1x1      8    double array
  the_answer_is            1x1      8    double array
  w                        1x1      8    double array
  x                        1x1      8    double array
  x1                       1x1      8    double array
  x2                       1x1      8    double array
  x3                       1x1      8    double array
  y                        1x1      8    double array
  z                        1x1      8    double array

Grand total is 16 elements using 128 bytes
```

To clear all the variables in a session, use the clear command as follows:

```
>> clear
```

After you use the clear command, the command who or whos will not display any variables. Here is another example. If you know the radius of a circle, then calculate both the perimeter and area of the circle. The solution is:

```
radius =

   12.3000
```

```
>> perimeter = 2*pi*radius

perimeter =

    77.2832

>> area = pi*radius^2

area =

   475.2916
```

In the above example, we have used the constant `pi` that was defined in Chapter 2. Also, note the use of the two arithmetic operations of multiplication and exponentiation.

Variables with the MATLAB Symbolic Math Toolbox

You may also use symbolic variables[6] in MATLAB. For example, the variables x and y are defined as symbolic variables as follows:

```
>> x=1/2

x =

    0.5000

>> y= 2/3

y =

    0.6667
```

---

[6] Actually these are symbolic objects in MATLAB but objects are beyond the scope of this book.

```
>> sym(x)

ans =

1/2


>> sym(y)

ans =

2/3
```

Arithmetic operations can now be performed on these symbolic variables. Consider the following example where the variables x and y were defined as above :

```
>> z=sym(x-y)

z =

-1/6
```

In the above example, the variable z is used automatically as a symbolic variable. For symbolic operations, the important thing is to start each operation or calculation with the sym command. Here is another example to calculate the volume of a sphere:

```
>> r=15/7

r =

    2.1429

>> sym(r)
```

```
ans =

15/7


>> volume = sym(4*pi*r^3/3)

volume =

4500*pi/343
```

In the above example, the final answer is given in the form of a fraction, $\dfrac{4500\pi}{343}$, with no decimal expressions and without calculating a numerical value (no units are used in this example). In case you need to have this value calculated numerically, then you can need to use the double command as follows:

```
>> double(volume)

ans =

    41.2162
```

It is seen thus that the value of the volume of the sphere is 41.2162 (no units are used in this example). Finally, the variables x and this can be defined as symbolic variables without assigning a numerical value as follows:

```
>> x = sym('x')

x =

x
```

```
>> this = sym('this')

this =

this
```

In the next chapter, we will study mathematical functions and their use with variables.

Exercises

Solve all the exercises using MATLAB. All the needed MATLAB commands for these exercises were presented in this chapter. Note that Exercises 21-27 require the use of the MATLAB Symbolic Math Toolbox.

1. Perform the operation $2*3+7$ and store the result in the variable w.

2. Define the three variables a, b, and c equal to 4, −10, and 3.2, respectively.

3. Define the two variables y and Y equal to 10 and 100. Are the two variables identical?

4. Let x $=$ 5.5 and y $=$ −2.6. Calculate the value of the variable z $=$ 2x−3y.

5. In Exercise 4 above, calculate the value of the variable w $=$ 3y − z + x/y.

6. Let r $=$ 6.3 and s $=$ 5.8. Calculate the value of the variable final defined by final $=$ r + s − r*s.

7. In Exercise 6 above, calculate the value of the variable
    `this_is_the_result`   defined   by
    `this_is_the_result = r^2 - s^2.`

8. Define the three variable `width`, `Width`, and `WIDTH` equal
    to 1.5, 2.0, and 4.5, respectively. Are these three variables
    identical?

9. Write the following comment in MATLAB: `This line`
    `will not be executed.`

10. Assign the value of `3.5` to the variable `s` then add a
    comment about this assignment on the same line.

11. Define the values of the variables `y1` and `y2` equal to 7 and
    9 then perform the calculation `y3 = y1 - y2/3.` (Note:
    2 in the formula is a subscript and should not be divided by
    3).

12. Perform the operation `2*m - 5.` Do you get an error?
    Why?

13. Define the variables `cost` and `profit` equal to `175` and
    `25`, respectively, then calculate the variable `sale_price`
    defined by `sale_price = cost + profit.`

14. Define the variable `centigrade` equal to `28` then calculate
    the variable `fahrenheit` defined by `fahrenheit =`
    `(centigrade*9/5) + 32.`

15. Use the `format short` and `format long` commands
    to write the values of `14/9` to four decimals and sixteen
    digits, respectively.

16. Perform the `who` command to get a list of the variables
    stored in this session.

17. Perform the `whos` command to get a list of the variables stored in this session along with their details.

18. Clear all the variables stored in this session by using the `clear` command.

19. Calculate both the area and perimeter of a rectangle of sides 5 and 7. No units are used in this exercise.

20. Calculate both the area and perimeter of a circle of radius 6.45. No units are used in this exercise.

21. Define the symbolic variables x and z with values 4/5 and 14/17.

22. In Exercise 21 above, calculate symbolically the value of the symbolic variable y defined by $y = 2x - z$.

23. Calculate symbolically the area of a circle of radius 2/3 without obtaining a numerical value. No units are used in this exercise.

24. Calculate symbolically the volume of a sphere of radius 2/3 without obtaining a numerical value. No units are used in this exercise.

25. In Exercise 23 above, use the `double` command to obtain the numerical value of the answer.

26. In Exercise 24 above, use the `double` command to obtain the numerical value of the answer.

27. Define the symbolic variables `y` and `date` without assigning any numerical values to them.

4. Mathematical Functions

In this chapter we introduce the use of mathematical functions in MATLAB. These functions include the square root function, the factorial, the trigonometric functions, the exponential functions, and the rounding and remainder functions[7]. We will start with the square root function[8] used with the MATLAB command `sqrt` as follows:

```
>> sqrt(2)

ans =

    1.4142
```

The above result gives the value of $\sqrt{2}$. The factorial of an integer is obtained using the MATLAB command `factorial` as follows:

```
>> factorial(5)

ans =

    120
```

All the trigonometric functions are also available within MATLAB. For example the sine and cosine functions are used with the MATLAB commands `sin` and `cos` as follows:

```
>> sin(30)
```

---

[7] MATLAB also has the special functions of mathematics like the Gamma function, the Error function, and Bessel functions but these are beyond the scope of this book.

[8] In this chapter we deal with square roots of positive numbers. Square roots of negative numbers are dealt with in the next chapter where we introduce complex numbers.

ans =

   -0.9880

>> sin(30*pi/180)

ans =

   0.5000

>> cos(30*pi/180)

ans =

   0.8660

Note that the argument for the trigonometric functions must be in radians. Check the first example above to see that the sine of the angle of 30 radians is -0.9880 but the sine of the angle of 30 degrees (converted into radians by multiplying it by $\frac{\pi}{180}$) is 0.5. The cosine of the angle of 30 degrees is seen above to be 0.8660. In addition, the tangent function is represented using the MATLAB command tan as follows where we compute the tangent of the angle of 30 degrees:

>> tan(30*pi/180)

ans =

   0.5774

The other trigonometric functions of secant, cosecant and cotangent are all represented in MATLAB using the commands sec, csc, and cot. In addition, the inverse trigonometric functions are represented by the MATLAB commands `asin`, `acos`, `atan`, `asec`, `acsc`, and `acot`. For example, the inverse tangent function is calculated as follows:

```
>> atan(1)

ans =

    0.7854
```

The angle returned in the above example is in radians. The hyperbolic trigonometric functions are represented by the MATLAB commands `sinh`, `cosh`, `tanh`, `sech`, `csch`, and `coth`. In addition, the inverse hyperbolic trigonometric functions are represented by the MATLAB commands `asinh`, `acosh`, `atanh`, `asech`, `acsch`, and `acoth`. Using these commands in MATLAB is straightforward. However, you need to consult a book on mathematics or trigonometry to check any limitations on the arguments of these functions. For example, if you try to find the tangent of the angle $\frac{\pi}{2}$ (90 degrees), then you will get a very large quantity as follows:

```
>> tan(pi/2)

ans =

    1.6331e+016
```

The reason for the strange result above is that the tangent function is not defined for the angle 90 degrees and the result given by MATLAB approaches infinity (thus we get a very large quantity).

The exponential and logarithmic functions are also represented by MATLAB. For example, the exponential function is represented by the MATLAB command `exp` as follows:

```
>> exp(1)

ans =

    2.7183
```

The above result gives the value of e^1. There are several logarithmic functions available in MATLAB. For example, the natural logarithm is represented by the command `log` and the logarithm to the base 10 is represented by the command `log10`. Here are two examples to illustrate this:

```
>> log(4)

ans =

    1.3863

>> log10(4)

ans =

    0.6021
```

The above two results give the values of $\ln 4$ and $\log_{10} 4$. We may also use the MATLAB command `log2` to obtain the logarithm to the base 2. For example, `log2(4)` will result in the value for $\log_2 4 = 2$.

Several rounding and remainder functions are also available in MATLAB. These functions are represented by the MATLAB commands `fix`, `floor`, `ceil`, `round`, `mod`, `rem`, and `sign`. For example, the rounding function `round` and the remainder function `rem` are used as follows:

```
>> round(4.57)

ans =

     5

>> rem(10,4)

ans =

     2
```

The above result gives the remainder when dividing 10 by 4. Finally, the absolute value function is represented by the MATLAB command abs as shown in the following example:

```
>> abs(-7)

ans =

     7
```

There are numerous other functions available in MATLAB like the complex functions which will be covered in the next chapter. Also, there are functions for coordinate transformations and there are number theoretic functions but these are beyond the scope of this book.

Mathematical functions can be combined using several commands on the same line in MATLAB and used with arithmetic operations. Here are some examples:

```
>> 2 + sqrt(5.3/3)

ans =

     3.3292
```

```
>> sin(cos(3.2))

ans =

    -0.8405

>> 2*log(2)/tan(pi/16)

ans =

    6.9694
```

The important point to consider when dealing with mathematical functions is to know what type of argument is valid for each function. For example, if you try to compute the natural logarithm of zero, then you will get an error or minus infinity because this function is not defined for zero. Here is the result:

```
>> log(0)
Warning: Log of zero.

ans =

   -Inf
```

Finally, mathematical functions can be used with variables. Here is an example:

```
>> a = 5

a =

      5

>> b = 2
```

```
b =

    2

>> c = b - sqrt(a+b/2)

c =

   -0.4495
```

Mathematical Functions with the MATLAB Symbolic Math Toolbox

Mathematical functions can also be used with the MATLAB Symbolic Math Toolbox. For example, the square root of 40 can be computed symbolically without performing any numerical calculations as follows:

```
>> sym(sqrt(40))

ans =

sqrt(40)


>> simplify(ans)

ans =

2*10^(1/2)
```

In the above, the `simplify` command is used to obtain the square root of 40 as $\sqrt{40} = 2\sqrt{10}$. To obtain a numerical result, use the `double` command applied on the previous result as follows:

```
>> double(ans)
```

```
ans =

    6.3246
```

The sine and cosine of the angle 30 degrees are computed symbolically as follows:

```
>> sym(sin(30*pi/180))

ans =

1/2
```

```
>> sym(cos(30*pi/180))

ans =

sqrt(3/4)
```

In addition, the tangent of the angle 30 degrees is obtained symbolically as follows:

```
>> sym(tan(30*pi/180))

ans =

sqrt(1/3)
```

Finally, the exponential and logarithm functions can also be used with Symbolic Math Toolbox. For example, the following command will compute $e^{2\pi}$ symbolically:

```
>> sym(exp(2*pi))
```

```
ans =

4710234414611992*2^(-43)
```

In order to obtain a numerical result, use the `double` command applied to the previous answer as follows:

```
>> double(ans)

ans =

   535.4917
```

In the next chapter, we will introduce complex numbers in MATLAB and how to handle them using complex functions

Exercises

Solve all the exercises using MATLAB. All the needed MATLAB commands for these exercises were presented in this chapter. Note that Exercises 19-25 require the use of the MATLAB Symbolic Math Toolbox.

1. Compute the square root of 10.

2. Compute the factorial of 7.

3. Compute the cosine of the angle 45 where 45 is in radians.

4. Compute the cosine of the angle 45 where 45 is in degrees.

5. Compute the sine of the angle of 45 where 45 is in degrees.

6. Compute the tangent of the angle 45 where 45 is in degrees.

7. Compute the inverse tangent of 1.5.

8. Compute the tangent of the angle $\dfrac{3\pi}{2}$. Do you get an error? Why?

9. Compute the value of exponential function e^3.

10. Compute the value of the natural logarithm $\ln 3.5$.

11. Compute the value of the logarithm $\log_{10} 3.5$.

12. Use the MATLAB rounding function round to round the value of 2.43.

13. Use the MATLAB remainder function rem to obtain the remainder when dividing 5 by 4.

14. Compute the absolute value of -3.6.

15. Compute the value of the expression $1.5 - 2\sqrt{6.7/5}$.

16. Compute the value of $\sin^2 \pi + \cos^2 \pi$.

17. Compute the value of $\log_{10} 0$. Do you get an error? Why?

18. Let $x = \dfrac{3\pi}{2}$ and $y = 2\pi$. Compute the value of the expression $2\sin x \cos y$.

19. Compute the value of $\sqrt{45}$ symbolically and simplify the result.

20. Compute the value of $\sqrt{45}$ numerically.

21. Compute the sine of the angle 45 (degrees) symbolically.

22. Compute the cosine of the angle 45 (degrees) symbolically.

23. Compute the tangent of the angle 45 (degrees) symbolically.

24. Compute the value of $e^{\pi/2}$ symbolically.

25. Compute the value of $e^{\pi/2}$ numerically.

5. Complex Numbers

In this chapter we discuss complex numbers and how to deal with them in MATLAB. A complex number has two parts – one real part and one imaginary part. The imaginary part of a complex number is usually the square root of a negative real number. The simplest complex number is the square root of -1 which is obtained in MATLAB as follows:

```
>> sqrt(-1)

ans =

        0 + 1.0000i
```

The imaginary part of a complex number is usually denoted by the symbol i where $i = \sqrt{-1}$. In this case the real part is zero. Some people prefer to use the symbol j instead of i to denote imaginary parts of complex numbers. Here is an example of a complex number:

```
>> 3+sqrt(-7)

ans =

   3.0000 + 2.6458i
```

In the above example, the real part is 3.0 and the imaginary part is 2.6458. We can also use variables with complex numbers. Define the variables a and b as complex numbers as follows:

```
>> a = 3+2i

a =
```

```
    3.0000 + 2.0000i
```

`>> b = 5-4i`

`b =`

```
    5.0000 - 4.0000i
```

Arithmetic operations can also be performed on complex numbers. For example, addition, subtraction, multiplication, and division can be performed on the variables a and b as follows:

`>> a+b`

`ans =`

```
    8.0000 - 2.0000i
```

`>> a-b`

`ans =`

```
   -2.0000 + 6.0000i
```

`>> a*b`

`ans =`

```
   23.0000 - 2.0000i
```

`>> a/b`

`ans =`

```
    0.1707 + 0.5366i
```

Exponentiation can also be performed on the complex numbers a and b as follows:

```
>> a^3

ans =

  -9.0000 +46.0000i

>> b^(-2)

ans =

   0.0054 + 0.0238i
```

Multiple arithmetic operations can also be performed on the complex numbers a and b as follows (the result is always a complex number):

```
>> 2*a + 3*b -5

ans =

  16.0000 - 8.0000i

>> a - i*b*4

ans =

 -13.0000 -18.0000i
```

There are some mathematical functions that can be used with complex numbers. For example, the MATLAB commands abs and angle can be used to obtain the magnitude and angle of a complex number, respectively. Here is an example of how to use these two commands:

```
>> abs(2+3i)
```

```
ans =

    3.6056

>> angle(2+3i)

ans =

    0.9828
```

In the above example, the angle of the complex number is obtained in radians. If you need the angle in degrees, then you have to multiply the result by $\dfrac{180}{\pi}$. Here is an example:

```
>> angle(2+3i)*180/pi

ans =

    56.3099
```

The real and imaginary parts of a complex number can be extracted using the MATLAB commands real and imag as follows:

```
>> real(2+3i)

ans =

        2

>> imag(2+3i)

ans =

        3
```

The complex conjugate of a complex number can be obtained using the MATLAB command `conj` as follows:
```
>> conj(2+3i)

ans =

   2.0000 - 3.0000i
```

A new complex number may be formed from real and imaginary parts using the MATLAB command `complex` as follows:

```
>> complex(2,3)

ans =

   2.0000 + 3.0000i
```

Trigonometric functions can also be used with complex numbers. Here are some examples:

```
>> sin(2+3i)

ans =

   9.1545 - 4.1689i

>> cos(2+3i)

ans =

  -4.1896 - 9.1092i

>> tan(2+3i)

ans =
```

```
-0.0038 + 1.0032i
```

In the above examples, the result is always a complex number. Exponential and logarithmic functions can also be used with complex numbers. Here are two examples:

```
>> exp(2+3i)

ans =

  -7.3151 + 1.0427i

>> log(2+3i)

ans =

   1.2825 + 0.9828i
```

In the above example, we have computed e^{2+3i} and $\ln(2+3i)$, respectively. Finally, here are some examples for computing the quantities $\sin(2\pi i)$, $\cos(2\pi i)$, and $e^{2\pi i}$:

```
>> sin(2*pi*i)

ans =

        0 +2.6774e+002i

>> cos(2*pi*i)

ans =

  267.7468

>> exp(2*pi*i)
```

```
ans =

 1.0  -  0.0000i
```

From the last example above, we note that the result of $e^{2\pi i}$ is the real number 1. We can also raise quantities to a complex number. Here is an example:

```
>> (2+3i)^(1-5i)

ans =

 3.2272e+002 +3.7004e+002i
```

As can be seen above, the result is usually a complex number.

Complex Numbers with the MATLAB Symbolic Math Toolbox

The MATLAB Symbolic Math Toolbox can also be used with complex numbers. For example, we define $\sqrt{-3}$ using the sym command as follows:

```
>> sym(sqrt(-3))

ans =

(0)+(sqrt(3))*i
```

The magnitude and angle of a complex number may be obtained symbolically using the abs and angle commands along with the sym command as follows:

```
>> sym(abs(2+3i))

ans =
```

```
sqrt(13)
```

```
>> sym(angle(2+3i))

ans =

8852218891597467*2^(-53)

>> double(ans)

ans =

    0.9828
```

In the above example, we used the double command to obtain the angle numerically in radians. Trigonometric functions can also be used with the sym command along with complex numbers but the double command may need to be used to obtain the final result numerically. Here is an example:

```
>> sym(sin(2+3i))

ans =

(5153524868349230*2^(-49))-
(4693771957861922*2^(-50))*i

>> double(ans)

ans =

    9.1545 - 4.1689i
```

The same remark above applies also to the exponential, logarithmic, and other mathematical functions in MATLAB. In the next chapter, we introduce vectors and their use in MATLAB.

Exercises

Solve all the exercises using MATLAB. All the needed MATLAB commands for these exercises were presented in this chapter. Note that Exercises 18-21 require the use of the MATLAB Symbolic Math Toolbox.

1. Compute the square root of -5.

2. Define the complex number $4 - 3\sqrt{-8}$.

3. Define the two complex numbers with variables x and y where $x = 2 - 6i$ and $y = 4 + 11i$.

4. In Exercise 3 above, perform the addition and subtraction operations $x + y$ and $x - y$.

5. In Exercise 3 above, perform the multiplication and division operations $x y$ and $\dfrac{x}{y}$.

6. In Exercise 3 above, perform the exponentiation operations x^4 and y^{-3} .

7. In Exercise 3 above, perform the multiple operations $4x - 3y + 9$.

8. In Exercise 3 above, perform the multiple operations $ix - 2y - 1$.

9. Compute the magnitude of the complex number $3 - 5i$.

10. Compute the angle of the complex number $3-5i$ in radians.

11. Compute the angle of the complex number $3-5i$ in degrees.

12. Extract the real and imaginary parts of the complex number $3-5i$.

13. Obtain the complex conjugate of the complex number $3-5i$.

14. Compute the sine, cosine, and tangent functions of the complex number $3-5i$.

15. Compute e^{3-5i} and $\ln(3-5i)$.

16. Compute the values of $\sin\dfrac{\pi i}{2}$, $\cos\dfrac{\pi i}{2}$, and $e^{\pi i/2}$.

17. Compute the value of $(3+4i)^{(2-i)}$.

18. Obtain $\sqrt{-13}$ symbolically.

19. Obtain the magnitude of the complex number $3-5i$ symbolically.

20. Obtain the angle of the complex number $3-5i$ symbolically. Make sure that you use the `double` command at the end.

21. Obtain the cosine function of the complex number $3-5i$ symbolically. Make sure that you use the `double` command at the end.

6. Vectors

In this chapter we introduce vectors[9] and how to deal with them in MATLAB. Vectors are stored in MATLAB as one-dimensional arrays. For example, here is a vector stored in the variable x:

```
>> x = [ 1 3 0 5 -2 ]

x =

     1     3     0     5    -2
```

In the above example, the vector x contains five elements. It is noticed that the elements of a vector in MATLAB are separated by spaces and enclosed between brackets. To extract the second element of the vector x, we use the notation x(2) as follows:

```
>> x(2)

ans =

     3
```

Here is another example where we extract the fifth element of the vector x:

```
>> x(5)

ans =

    -2
```

---

[9] We mean here vectors of numbers or scalars. There are also vectors of strings and characters in MATLAB but these are beyond the scope of this book.

Here is a vector y defined implicitly with five elements as follows:

```
>> y = pi*x

y =

    3.1416   9.4248      0   15.7080   -6.2832
```

In the above example, the vector y is obtained from the vector x by multiplying each element of the vector x by π. To extract the fourth element of the vector y, we use the notation y(4) as follows:

```
>> y(4)

ans =

   15.7080
```

As we have seen above, we use the notation y(n) where y is the variable where the vector is stored and n is an integer ranging from 1 to the length of the vector. We have to be careful when we use this notation – n has to be an integer. If we try to use y(1.2) then we obtain an error message as follows (because 1.2 is not an integer):

```
>> y(1.2)
??? Subscript indices must either be real
positive integers or logicals.
```

In order to extract the first three elements of the vector y, we use the notation y(1:3) as follows:

```
>> y(1:3)

ans =
```

```
    3.1416      9.4248                0
```

The length of a vector is the number of elements contained in the vector. In order to obtain the length of a vector, we use the MATLAB command `length` as follows:

```
>> length(y)

ans =

    5
```

In order to add the total sum of the values of the elements of a vector, we use the MATLAB command `sum` as follows:

```
>> sum(y)

ans =

   21.9911
```

In order to find the minimum value and the maximum value of the elements of a vector, we use the MATLAB commands `min` and `max`, respectively, as follows:

```
>> min(y)

ans =

   -6.2832

>> max(y)

ans =

   15.7080
```

In the above examples of vectors, we constructed the vectors by listing their elements either explicitly or implicitly. There are two other ways to construct vectors in MATLAB. For example, the vector a is constructed (using parentheses) as follows:

```
>> a = (0:0.2:1)

a =

        0      0.2000      0.4000      0.6000
0.8000    1.0000
```

In the above example, the vector a contains all the real numbers between 0 and 1 with an increment of 0.2. Another way to construct vectors in MATLAB is to use the linspace[10] command (using parentheses) as follows:

```
>> b = linspace(0,5,10)

b =

  Columns 1 through 6

        0      0.5556      1.1111      1.6667
2.2222    2.7778

  Columns 7 through 10

    3.3333      3.8889      4.4444      5.0000
```

In the above example, the vector b is constructed such that it contains 10 elements between 0 and 5 that are equally spaced[11]. Ignore the MATLAB output where it says "Columns ? to ?" –

---

[10] There is also a logspace command in MATLAB but its use is beyond the scope of this book. It basically deals with generating logarithmically spaced vectors.
[11] Actually, to be precise, it is called linearly spaced.

this notation will become clear to us when we study matrices in the next chapter.

Two vectors can be joined together to form a new vector. In the following example, the two vectors a and b are joined together to form the new vector c:

```
>> a = [ 1 3 5]

a =

        1        3        5

>> b = [ 7 9 ]

b =

        7        9

>> c = [a b]

c =

        1        3        5        7        9
```

Based on the above example, we can also form a new vector d as follows:

```
>> d = [a b 11]

d =

        1        3        5        7        9        11
```

Next, we will discuss the arithmetic operations that are allowed with vectors. The operations of addition and subtraction can be performed on two vectors of the same length. Here is an example:

```
>> x = [0.1 0.3 -2 5]

x =

    0.1000      0.3000     -2.0000      5.0000

>> y = [0.25 0.4 1 0]

y =

    0.2500      0.4000      1.0000           0

>> x+y

ans =

    0.3500      0.7000     -1.0000      5.0000

>> x-y

ans =

   -0.1500     -0.1000     -3.0000      5.0000
```

If you try to add or subtract two vectors of different lengths, then you will get an error message. Here is an example:

```
>> a = [1 5 3 5 6]

a =

     1       5       3       5       6

>> b = [-2 3 5 7]

b =
```

```
    -2      3      5      7
```

```
>> a+b
??? Error using ==> plus
Matrix dimensions must agree.
```

```
>> a-b
??? Error using ==> minus
Matrix dimensions must agree.
```

The operation of multiplication is a little more complicated for vectors. Even if you try to multiply two vectors of the same length, you will get an error message. Here is an example using the vectors x and y of the same length defined above:

```
>> x*y
??? Error using ==> mtimes
Inner matrix dimensions must agree.
```

Note that the operation of division is absolutely not defined for vectors – you cannot divide one vector by another. However, you can perform the new operations of element-by-element multiplication and element-by-element division of vectors in MATLAB using the vectors x and y defined above as follows:

```
>> x.*y

ans =

    0.0250    0.1200    -2.0000              0

>> x./y
Warning: Divide by zero.

ans =
```

```
    0.4000      0.7500     -2.0000             Inf
```

Note in the above example that the symbols for multiplication and division are preceded by a dot. The dot tells MATLAB to perform the operations element by element. Note that in order to carry these element-by-element operations, the two vectors must be of the same length.

There is another type of multiplication that is defined for vectors. It is called the scalar product, the inner product, or the dot product[12] – three different names for the same operation. This product is obtained in MATLAB for the two vectors x and y defined above as follows:

```
>> x*y'
```

```
ans =
```

```
    -1.8550
```

The final outcome of this operation is always a scalar or a number – not a vector. In addition, the two vectors that are used must be of the same length. Note the use of the prime on the second vector in the operation above. The use of the prime indicates the transpose operation which we will discussed in detail in the next chapter on matrices.

There are other simpler operations that can be performed on vectors. The operations of scalar addition and scalar subtraction can be performed as in the following example:

```
> s = [1 4 -3 2 2]
```

```
s =
```

---

[12] The reader is advised to consult a book on vector analysis for more details about this product.

      1   4   −3   2   2

```
>> s+4
```

ans =

      5   8   1   6   6

```
>> s-3
```

ans =

    −2   1   −6   −1   −1

In the above example, the number 4 is added to each element of the vector while the number 3 is subtracted from each element of the vector. In addition, the operations of scalar multiplication and scalar division can also be performed on the vector s defined above as follows:

```
>> s*2
```

ans =

      2   8   −6   4   4

```
>> s/5
```

ans =

    0.2000   0.8000   −0.6000   0.4000
0.4000

In the first example above, each element of the vector s is multiplied by the number 2 while in the second example, each element of the vector s is divided by the number 5. The above four operations can also be combined together using the vector s defined above as follows:

```
>> 2+ s/2.4
```

ans =

   2.4167   3.6667   0.7500   2.8333   2.8333

Mathematical functions can also be used with vectors. Here is an example where we use the three trigonometric functions on the vector a defined below:

```
>> a = [0 pi/2 pi 3*pi/2 2*pi]
```

a =

        0   1.5708   3.1416   4.7124
6.2832

```
>> sin(a)
```

ans =

        0   1.0000   0.0000   -1.0000   -
0.0000

```
>> cos(a)
```

ans =

    1.0000   0.0000   -1.0000   -0.0000
1.0000

```
>> tan(a)
```

ans =

  1.0e+016 *

        0   1.6331   -0.0000   0.5444   -
0.0000

In the above example, the trigonometric operations are carried out element-by-element. Here is an example with the same vector a where we use the exponential function:

>> exp(a)

ans =

    1.0000   4.8105   23.1407   111.3178
535.4917

In the above example, the exponential function is carried out element-by-element. Here is another example where we use the square root functions on the vector a defined above as follows:

>> sqrt(a)

ans =

        0   1.2533   1.7725   2.1708
2.5066

Note that the square root operation was performed element by element in the example above. The last operation that we will discuss is the exponentiation operation. If we try to evaluate the quantity 2^a where a is the vector defined above, then we will receive an error message as follows:

>> 2^a

```
??? Error using ==> mpower
Matrix must be square.
```

In order to carry out the exponentiation operation element by element, we need to include the dot symbol before the exponentiation symbol as follows:

```
>> 2.^a

ans =

    1.0000        2.9707        8.8250       26.2162
77.8802
```

In the above example, the exponentiation operation is carried out element by element.

There are certain types of vectors that can be generated automatically by MATLAB. These types of vectors have specific commands for their generation. For example, the MATLAB commands ones and zeros produce vectors with elements of 1's and 0's, respectively. Here are two examples:

```
>> t = ones(1,7)

t =

    1     1     1     1     1     1     1

>> w = zeros(1,5)

w =

    0     0     0     0     0
```

The exact use of the above two commands will be discussed in detail in the next chapter on matrices. Elements of a vector can be sorted using the MATLAB command `sort` as follows:

```
>> x = [2 1 -3 5 3]

x =

    2      1     -3      5      3

>> sort(x)

ans =

   -3      1      2      3      5
```

Sorting in the above example has been obtained in an ascending order[13]. Finally, a vector with its element forming a random permutation can be generated using the `randperm` command as follows:

```
>> randperm(7)

ans =

    2      7      4      3      6      5      1
```

In the above example, a random permutation of seven elements is obtained.

There are some MATLAB commands for statistics that can be used with vectors. For example, the MATLAB commands `range`, `mean`, and `median` can be used to obtain the statistical

---

[13] Sorting can also be obtained in a descending order but this is beyond the scope of this book. This simple example of sorting is included to show the various capabilities of MATLAB.

values of range, mean, and median for a set of numbers in a vector as follows:

```
>> w = [1 5 -2 0 4 6]

w =

      1      5     -2      0      4      6

>> range(w)

ans =

      8

>> mean(w)

ans =

    2.3333

>> median(w)

ans =

    2.5000
```

Vectors with the MATLAB Symbolic Math Toolbox

In this section, we will discuss the use of symbolic vectors in MATLAB. Symbolic vectors are vectors whose elements are handled algebraically without numerical computations. In the following example, we define a symbolic vector a of three elements using the MATLAB command syms:

```
>> syms x y z
>> a = [x y z]

a =

[ x, y, z]
```

Next, we carry out the following scalar addition operation on the vector a defined previously to obtain the new symbolic vector b:

```
>> b = 2+a

b =

[ x+2, y+2, z+2]
```

Next, we extract the second element of the vector b as follows:

```
>> b(2)

ans =

y+2
```

Next, we perform the addition of the two vectors a and b to get the new symbolic vector c as follows:

```
>> c = a+b

c =

[ 2*x+2, 2*y+2, 2*z+2]
```

Next, we perform the operation $2*a - 3*b/5$ to obtain the new symbolic vector d as follows:

```
>> d = 2*a - 3*b/5

d =

[ 7/5*x-6/5, 7/5*y-6/5, 7/5*z-6/5]
```

Next, we obtain the dot product of the vector a and b as follows:

```
>> a*b'

ans =

x*(2+conj(x))+y*(2+conj(y))+z*(2+conj(z))
```

In the above example, notice that the result is a scalar quantity (a symbolic or algebraic expression) – not a vector. We can also perform the dot product in the following way by reversing the order of the two vectors:

```
>> b*a'

ans =

(x+2)*conj(x)+(y+2)*conj(y)+(z+2)*conj(z)
```

Notice that the two dot products obtained above are different. They would be exactly the same if the variables x, y, and z were real variables. MATLAB assumes by default that these are complex variables thus obtaining two different results.

Next, we obtain the square root of the vector a−b element by element as follows:

```
>> sqrt(a-b)
```

```
ans =
```

```
[ i*2^(1/2), i*2^(1/2), i*2^(1/2)]
```

The symbolic vector obtained above is a complex vector, i.e. a vector with complex elements. Finally, we can obtain a numerical value for the above expression by using the MATLAB command double as follows:

```
>> double(ans)
```

```
ans =
```

```
    0 + 1.4142i    0 + 1.4142i    0 + 1.4142i
```

It is clear from the above result that the final vector is a complex vector with its elements being complex numbers. In the next chapter we will discuss the use of matrices in MATLAB.

Exercises

Solve all the exercises using MATLAB. All the needed MATLAB commands for these exercises were presented in this chapter. Note that Exercises 36-41 require the use of the MATLAB Symbolic Math Toolbox.

1. Store the vector [2 4 -6 0] in the variable w.

2. In Exercise 1 above, extract the second element of the vector w.

3. In Exercise 1 above, generate the vector z where $z = \dfrac{\pi}{2} w$.

4. In Exercise 3 above, extract the fourth element of the vector z.

5. In Exercise 3 above, extract the first three elements of the vector z.

6. In Exercise 3 above, find the length of the vector z.

7. In Exercise 3 above, find the total sum of the values of the elements of the vector z.

8. In Exercise 3 above, find the minimum and maximum values of the elements of the vector z.

9. Generate a vector r with real values between 1 and 10 with an increment of 2.5.

10. Generate a vector s with real values of ten numbers that are equally spaced between 1 and 100.

11. Form a new vector by joining the two vectors [9 3 -2 5 0] and [1 2 -4].

12. Form a new vector by joining the vector [9 3 -2 5 0] with the number 4.

13. Add the two vectors [0.2 1.3 -3.5] and [0.5 -2.5 1.0].

14. Subtract the two vectors in Exercise 13 above.

15. Try to multiply the two vectors in Exercise 13 above. Do you get an error message? Why?

16. Multiply the two elements in Exercise 13 above element by element.

17. Divide the two elements in Exercise 13 above element by element.

18. Find the dot product of the two vectors in Exercise 13 above.

19. Try to add the two vectors [1 3 5] and [3 6]. Do you get an error message? Why?

20. Try to subtract the two vectors in Exercise 20 above. Do you get an error message? Why?

21. Let the vector w be defined by w = [0.1 1.3 -2.4]. Perform the operation of scalar addition 5+w.

22. In Exercise 22 above, perform the operation of scalar subtraction -2-w.

23. In Exercise 22 above, perform the operation of scalar multiplication 1.5*w.

24. In Exercise 22 above, perform the operation of scalar division w/10.

25. In Exercise 22 above, perform the operation 3 - 2*w/5.

26. Define the vector b by b = [0 pi/3 2pi/3 pi]. Evaluate the three vectors $\sin b$, $\cos b$, and $\tan b$ (element by element).

27. In Exercise 26 above, evaluate the vector e^b (element by element).

28. In Exercise 26 above, evaluate the vector $\sqrt{b}$ (element by element).

29. Try to evaluate the vector 3^b. Do you get an error message? Why?

30. Perform the operation in Exercise 29 above element by element?

31. Generate a vector of 1's with a length of 4 elements.

32. Generate a vector of 0's with a length of 6 elements.

33. Sort the elements of the vector [0.35 -1.0 0.24 1.30 -0.03] in ascending order.

34. Generate a random permutation vector with 5 elements.

35. For the vector [2 4 -3 0 1 5 7], determine the range, mean, and median.

36. Define the symbolic vector x = [r s t u v].

37. In Exercise 36 above, perform the addition operation of the two vectors x and [1 0 -2 3 5] to obtain the new symbolic vector y.

38. In Exercise 37 above, extract the third element of the vector y.

39. In Exercise 37 above, perform the operation 2*x/7 + 3*y.

40. In Exercise 37 above, perform the dot products x*y' and y*x'. Are the two results the same? Why?

41. In Exercise 37 above, find the square root of the symbolic vector x+y.

7. Matrices

In this chapter we introduce matrices and how to deal with them in MATLAB. Matrices are stored as two-dimensional arrays[14] in MATLAB. A matrix is a collection of numbers and/or scalars arranged in rows and columns. For example here is a matrix of numbers with three rows and four columns:

```
>> x = [2 1 -3 0 ; -2 3 4 1 ; 5 -2 1 3]

x =

     2     1    -3     0
    -2     3     4     1
     5    -2     1     3
```

In the above example, the matrix x is rectangular because the number of columns and the number of rows are not equal. In the specification of the matrix in MATLAB, the elements of the same row are separated by spaces while the rows themselves are separated by semicolons. Brackets are used to indicate the beginning and end of a matrix. In particular, the above matrix is called a 3x4 rectangular matrix.

We may extract the element in the second row and the third column by using the notation x(2,3) as follows:

```
>> x(2,3)

ans =

     4
```

---

[14] MATLAB can handle also multi-dimensional arrays but this is beyond the scope of this book.

We may also extract the element in the first row and the fourth column as follows

```
>> x(1,4)

ans =

    0
```

A new matrix of the same size (i.e the same number of rows and columns) may be generated by multiplying the matrix by a number as follows:

```
>> y = 2*pi*x

y =

    12.5664      6.2832   -18.8496          0
   -12.5664     18.8496    25.1327     6.2832
    31.4159    -12.5664     6.2832    18.8496
```

We may extract the element in the third row and the third column of the matrix y as follows:

```
>> y(3,3)

ans =

    6.2832
```

We may also extract the elements common in the first and second rows and the second and third columns of the matrix y as follows:

```
>> y(1:2,2:3)

ans =
```

```
    6.2832   -18.8496
   18.8496    25.1327
```

In the above example, we obtain a sub-matrix of size 2x2, i.e. with two rows and two columns. The size of a matrix is the number of rows and columns of the matrix. This is determined in MATLAB by using the command `size` as follows:

```
>> size(y)

ans =

     3        4
```

In the above example, we obtain two numbers – the first one is the number of rows while the second one is the number of columns. The MATLAB command `length` may also be used but it gives one number – the largest of the number of rows and the number of columns. Here is an example:

```
>> length(y)

ans =

     4
```

In the above example, we obtain the single number 4, i.e. the largest of 3 and 4. The number of elements in a matrix is obtained by using the MATLAB command `numel` as follows:

```
>> numel(y)

ans =

    12
```

In the above example, we obtain 12 which is the product of 3 and 4 – the total number of elements in the matrix. The MATLAB commands sum, min, and max may also be used as follows:

```
>> sum(y)

ans =

   31.4159    12.5664    12.5664    25.1327

>> min(y)

ans =

  -12.5664   -12.5664   -18.8496          0

>> max(y)

ans =

   31.4159    18.8496    25.1327    18.8496
```

In the above example, the MATLAB command sum gives the total sum of each column of the matrix, the MATLAB command min gives the minimum value of each column of the matrix, while the MATLAB command max gives the maximum value of each column of the matrix.

We can combine two or more matrices to obtain a new matrix. Here is an example of how to combine two vectors to obtain a matrix:

```
>> a = [1 3 4]

a =

     1     3     4
```

```
>> b = [5 6 7]

b =

      5        6        7

>> c = [a ; b]

c =

      1        3        4
      5        6        7
```

We can perform certain operation on matrices. For example, we can perform the operations of matrix addition and matrix subtraction on two matrices of the same size. Here is an example:

```
>> x = [0.2 -0.1 ; 1.3 0.0 ; 2.4 -1.5]

x =

     0.2000     -0.1000
     1.3000           0
     2.4000     -1.5000

>> y = [0.5 1.0 ; 1.2 1.3 ; -2.0 1.3]

y =

     0.5000      1.0000
     1.2000      1.3000
    -2.0000      1.3000

>> x+y

ans =
```

```
      0.7000      0.9000
      2.5000      1.3000
      0.4000     -0.2000

>> x-y

ans =

     -0.3000     -1.1000
      0.1000     -1.3000
      4.4000     -2.8000
```

In the above example, we added and subtracted two rectangular matrices of size 3x2. You will get an error message if you try to add or subtract two matrices of different sizes.

Matrix multiplication is not defined for two matrices of the same size (unless they are square matrices). You will get an error message if you try to multiply two matrices of the same size. Here is an example:

```
>> x*y
??? Error using ==> mtimes
Inner matrix dimensions must agree.
```

However, you can multiply two matrices of the same size element-by-element by adding the dot symbol before the multiplication symbol as follows:

```
>> x.*y

ans =

      0.1000     -0.1000
      1.5600           0
     -4.8000     -1.9500
```

The resulting matrix above is of the same size as the two matrices that were multiplied. Division of matrices is absolutely not defined mathematically[15]. You will get an error message if you try to divide two matrices of any size. However, you may divide two matrices of the same size element-by-element by including the dot symbol before the division symbol as follows:

```
>> x./y

ans =

    0.4000   -0.1000
    1.0833        0
   -1.2000   -1.1538
```

The resulting matrix above is of the same size as the two matrices that were divided. Next, we discuss the multiplication of matrices. Two matrices may be multiplied if the number of columns of the first matrix equals the number of rows of the second matrix. Here is an example:

```
>> u = [1 3 ; -2 0 ; 1 5]

u =

    1    3
   -2    0
    1    5

>> v = [3 4 1 -2 ; 1 0 -1 2]
```

---

[15] We can divide two matrices in MATLAB using the backslash operator "\" but this is not a true division operation but related to the solution of the system of equations represented by the matrices. For more details about this operation, see Chapter 10 on solving equations.

```
v  =
```

```
       3       4       1      -2
       1       0      -1       2
```

```
>> u*v
```

```
ans  =
```

```
       6       4      -2       4
      -6      -8      -2       4
       8       4      -4       8
```

In the above example, we multiplied a 3x2 matrix by a 2x4 matrix to obtain a 3x4 matrix. It is clear that the rule of matrix multiplication applies in this case.

Next, we discuss the operations of scalar addition, scalar subtraction, scalar multiplication, and scalar division. A scalar can be added or subtracted from a matrix as follows:

```
>> s = [1 3 -2 5 ; 2 6 -3 0]
```

```
s  =
```

```
       1       3      -2       5
       2       6      -3       0
```

```
>> s+5
```

```
ans  =
```

```
       6       8       3      10
       7      11       2       5
```

```
>> s-2
```

```
ans =

    -1      1     -4      3
     0      4     -5     -2
```

    In the above scalar addition and scalar subtraction operations, the number 5 was added to each element of the matrix s, while the number 2 was subtracted from each element of the matrix s. Scalar multiplication and scalar division may be performed in the same way as follows:

```
>> s*3

ans =

     3      9     -6     15
     6     18     -9      0

>> s/4

ans =

    0.2500    0.7500   -0.5000    1.2500
    0.5000    1.5000   -0.7500        0
```

    In the above example, the number 3 was multiplied by each element of the matrix s, while the each element of the matrix s was divided by the number 4. In addition, multiple scalar operations may be performed on the same line as follows:

```
>> 3 - 2*s/1.5

ans =

    1.6667   -1.0000    5.6667   -3.6667
    0.3333   -5.0000    7.0000    3.0000
```

Mathematical functions can also be used with matrices. In the following example, we use the trigonometric functions of sine, cosine, and tangent. All the calculations are performed element-by-element using the MATLAB commands sin, cos, and tan as follows:

```
>> w = [0 pi/2 pi ; pi/2 pi 3*pi/2 ; pi
3*pi/2 2*pi]

w =

        0      1.5708      3.1416
   1.5708      3.1416      4.7124
   3.1416      4.7124      6.2832

>> sin(w)

ans =

        0      1.0000      0.0000
   1.0000      0.0000     -1.0000
   0.0000     -1.0000     -0.0000

>> cos(w)

ans =

   1.0000      0.0000     -1.0000
   0.0000     -1.0000     -0.0000
  -1.0000     -0.0000      1.0000

>> tan(w)

ans =

   1.0e+016 *
```

```
        0       1.6331      -0.0000
   1.6331      -0.0000       0.5444
  -0.0000       0.5444      -0.0000
```

The square root of a matrix can also be calculated in MATLAB using two different approaches. In the first approach, the square root is calculated using the usual MATLAB command sqrt with the computations performed element-by-element as follows:

```
>> sqrt(w)
```

ans =

```
        0       1.2533       1.7725
   1.2533       1.7725       2.1708
   1.7725       2.1708       2.5066
```

However, in the above example we do not obtain a true square root of the matrix. In order to find the true square root of the matrix, we need to use the special matrix MATLAB command sqrtm as follows:

```
>> sqrtm(w)
```

ans =

```
   0.3086   +   0.8659i        0.5476   +   0.1952i
0.7867  -  0.4755i
   0.5476   +   0.1952i        0.9718   +   0.0440i
1.3960  -  0.1072i
   0.7867   -   0.4755i        1.3960   -   0.1072i
2.0053 + 0.2612i
```

Obviously, the result above is a complex matrix. If you try to multiply it by itself, you will obtain the original matrix. Similarly, there are two MATLAB commands for computing the exponential of a matrix. The first one is the usual command exp which performs the calculations element-by-element as follows:

```
>> exp(w)

ans =

    1.0000      4.8105     23.1407
    4.8105     23.1407    111.3178
   23.1407    111.3178    535.4917
```

However, in order to obtain a true exponential of a matrix, the special command of expm should be used as follows:

```
>> expm(w)

ans =

  1.0e+004 *

    0.4586      0.8138      1.1690
    0.8138      1.4442      2.0745
    1.1690      2.0745      2.9801
```

Similarly, there are two MATLAB commands for computing the natural logarithm of a matrix. The first one is the usual command of log which performs the calculations element-by-element as follows:

```
>> log(w)
Warning: Log of zero.

ans =

     -Inf      0.4516      1.1447
    0.4516      1.1447      1.5502
    1.1447      1.5502      1.8379
```

Note that the first element in the resulting matrix above is minus infinity – this is because we are trying to compute the natural logarithm of zero which is not defined. However, in order to obtain a true natural logarithm of a matrix, the special command of logm should be used as follows:

```
>> logm(w)
Warning: Principal  matrix  logarithm  is  not
defined for A with
          nonpositive   real   eigenvalues.   A
non-principal matrix
          logarithm is returned.
> In funm at 153
   In logm at 27

ans =

  -5.9054 + 2.8465i    13.1738 - 0.5236i    -
5.9209 - 0.7521i
   13.1738  -  0.5236i  -24.7338  +  2.2124i
13.7066 - 1.3348i
  -5.9209 - 0.7521i    13.7066 - 1.3348i    -
4.8399 + 1.2242i
```

Obviously, a complex matrix is generated as the true natural logarithm. Note the remark generated by MATLAB which says that the natural logarithm that is generated is non-principal. We can also perform the following exponential operation:

```
>> 2^w

ans =

  1.0e+003 *

   0.1674    0.2961    0.4257
```

```
    0.2961      0.5266      0.7551
    0.4257      0.7551      1.0854
```

The above exponential operation may also be performed element-by-element by including the dot symbol as follows:

```
>> 2.^w
```

ans =

```
    1.0000      2.9707      8.8250
    2.9707      8.8250     26.2162
    8.8250     26.2162     77.8802
```

Other operations may also be performed on the matrix w. For example, the cube of the matrix w, which is w^3 may be obtained as follows:

```
>> w^3
```

ans =

```
  116.2735    209.2924    302.3112
  209.2924    372.0753    534.8583
  302.3112    534.8583    767.4053
```

There are some standard matrices that can be generated automatically by MATLAB. For example, matrices with elements of 1's and 0's can be generated using the commands ones and zeros as follows:

```
>> ones(3,5)
```

ans =

```
    1       1       1       1       1
```

```
    1       1       1       1       1
    1       1       1       1       1
```

```
>> zeros(3,5)
```

```
ans =
```

```
    0       0       0       0       0
    0       0       0       0       0
    0       0       0       0       0
```

In addition, a special matrix called the identity matrix[16] can be generated using the MATLAB command eye as follows:

```
>> eye(3,5)
```

```
ans =
```

```
    1       0       0       0       0
    0       1       0       0       0
    0       0       1       0       0
```

A matrix with the same number of rows and columns is called a square matrix. For example, a matrix with three rows and three columns is called a square matrix of size 3. Here is an example:

```
>> m = [1 4 -2 ; 0 4 7 ; 1 -2 5]
```

```
m =
```

```
    1       4      -2
    0       4       7
    1      -2       5
```

---

[16] An identity matrix is a matrix with 1's on the main diagonal and 0's everywhere else. The identity matrix is very useful in matrix algebra.

For the remaining part of this discussion, we will deal with operations and handling of square matrices. For example, the above three commands of `ones`, `zeros`, and `eye` can be used as follows to generated standard square matrices as follows:

```
>> ones(3)

ans =

     1        1        1
     1        1        1
     1        1        1

>> zeros(3)

ans =

     0        0        0
     0        0        0
     0        0        0

>> eye(3)

ans =

     1        0        0
     0        1        0
     0        0        1
```

Notice the different usage of the above three commands when used to generate square matrices. The operations and commands that we will discuss next are part of the subject called matrix algebra. We will not provide a comprehensive coverage of this subject here, but we will provide the essential commands. The transpose of a matrix is generated in MATLAB using the prime symbol as follows:

```
>> A = [ 1 3 6 ; -2 5 0 ; 1 2 6]

A =

     1        3        6
    -2        5        0
     1        2        6

>> A'

ans =

     1       -2        1
     3        5        2
     6        0        6
```

In the above example, A' is the transpose[17] of A. It is generated by replacing each row in A with its corresponding column. In effect, the rows and columns are switched. For example, we next add the matrix A with its transpose:

```
>> A+A'

ans =

     2        1        7
     1       10        2
     7        2       12
```

Note that the resulting matrix is symmetric. The diagonal of a square matrix can be extracted using the MATLAB command diag as follows:

```
>> diag(A)
```

---

[17] The sum of a matrix and its transpose will always be a symmetric matrix.

```
ans =

    1
    5
    6
```

We can also extract the upper triangular part and the lower triangular part[18] of the matrix using the MATLAB commands triu and tril as follows:

```
>> triu(A)

ans =

    1        3        6
    0        5        0
    0        0        6

>> tril(A)

ans =

    1        0        0
   -2        5        0
    1        2        6
```

The determinant and inverse of a matrix are obtained by using the MATLAB commands det and inv as follows:

```
>> det(A)

ans =

    12
```

---

[18] Consult a book on matrix algebra to find the exact definition of the upper triangular part and the lower triangular part of a matrix.

```
>> inv(A)

ans =

    2.5000   -0.5000   -2.5000
    1.0000         0   -1.0000
   -0.7500    0.0833    0.9167
```

Note that the determinant is always a scalar (a number in this case) while the inverse of a square matrix is also a square matrix of the same size. If we multiply the matrix by its inverse matrix, we should obtain the identity matrix of the same size as follows:

```
>> A*inv(A)

ans =

    1.0000   -0.0000         0
   -0.0000    1.0000    0.0000
   -0.0000   -0.0000    1.0000
```

We can obtain the trace of a matrix using the MATLAB command trace as follows:

```
>> trace(A)

ans =

    12
```

The trace[19] of a matrix is the sum of the diagonal elements. It is always a scalar (a number in this case). The norm[20] of a matrix can also be computed using the MATLAB command norm as follows:

---

[19] The trace and determinant of a matrix are not equal although in our example here both turn out to be 12 – a coincidence.

```
>> norm(A)

ans =

    9.4988
```

Note from the example above that the norm of a matrix is always a scalar (a number in this case). We can also obtain the eigenvalues[21] of the matrix A using the MATLAB command eig as follows:

```
>> eig(A)

ans =

    0.3223
    5.8389 + 1.7732i
    5.8389 - 1.7732i
```

Note from the example above that we obtained three eigenvalues because the size of the matrix A is 3. Note also that some of the eigenvalues are complex numbers. We can also obtain the characteristic polynomial[22] of the matrix A using the MATLAB command poly as follows:

```
>> poly(A)

ans =

    1.0000   -12.0000    41.0000   -12.0000
```

---

[20] Consult a book on matrix algebra for the exact definition of the norm of a matrix.

[21] The eigenvalues of a matrix are its principal values.

[22] For details about the definitions of the eigenvalues of a matrix and its characteristic polynomial, consult a book on linear algebra.

Notice in the above result that we obtained the coefficients of the characteristic polynomial of the matrix A. Because A is a square matrix of size 3, its characteristic polynomial is cubic. Thus we obtain four coefficients for a cubic polynomial. We can obtain the rank of a matrix using the MATLAB command rank as follows:

```
>> rank(A)

ans =

    3
```

For a square matrix, the rank is the same as the size of the matrix. Finally, we will discuss briefly random matrices and magic matrices. To generate a random matrix (a matrix with random elements), we use the MATLAB command rand as follows:

```
>> rand(3)

ans =

    0.9501    0.4860    0.4565
    0.2311    0.8913    0.0185
    0.6068    0.7621    0.8214
```

As seen from the above example, the random elements of the matrix have values between 0 and 1. Of course, these values can be manipulated using the different matrix operations discussed previously in this chapter. And to generate a magic matrix (called a magic square), we use the MATLAB command magic as follows:

```
>> magic(3)

ans =

    8    1    6
```

```
       3        5        7
       4        9        2
```

The above magic square of size 3x3 has the feature that the sum of any column or row or diagonal is always the same number – in this example the sum is 15.

Matrices with the MATLAB Symbolic Math Toolbox

In this section we introduce symbolic matrices, i.e. matrices with symbolic variables that can be handled algebraically without performing any numerical computations. Most of the operations and commands discussed previously in this chapter can be used with symbolic matrices. Here is an example of a symbolic square matrix of size 3 defined using the MATLAB command syms:

```
>> syms x
>> A = [x  x-3  x^2 ;  2*x  1-x  x ;  3*x  2*x-5
x^2]

A =

[      x,      x-3,      x^2]
[    2*x,      1-x,        x]
[    3*x,    2*x-5,      x^2]
```

Here is another symbolic matrix based on the same symbolic variable x:

```
>> B = [ 1 0 x ;  -2 5 1-x ;  2 3 4]

B =

[     1,      0,      x]
[    -2,      5,    1-x]
```

```
[    2,    3,    4]
```

Next, we sum the two symbolic matrices A and B to obtain the new symbolic matrix C as follows:

```
>> C = A+B

C =

[    x+1,    x-3,  x^2+x]
[ 2*x-2,    6-x,      1]
[ 3*x+2,  2*x-2,  x^2+4]
```

The transpose of the symbolic matrix C is obtained using the prime symbol as follows:

```
>> C'

ans =

[    1+conj(x),  -2+2*conj(x),   2+3*conj(x)]
[   -3+conj(x),    6-conj(x),  -2+2*conj(x)]
[  conj(x^2+x),            1,  4+conj(x)^2]
```

The determinant of the symbolic matrix C is obtained using the MATLAB command det as follows:

```
>> det(C)

ans =

-4*x^3+37*x+4*x^4-4-43*x^2
```

The trace of the symbolic matrix C is obtained using the MATLAB command trace as follows:

```
>> trace(C)
```

```
ans =

11+x^2
```

Finally, the inverse of the symbolic matrix C can be obtained using the MATLAB command inv as follows:

```
>> inv(C)

ans =

[      -(-6*x^2-26+x^3+6*x)/(-4*x^3+37*x+4*x^4-
4-43*x^2),                    (x^3+3*x^2-6*x+12)/(-
4*x^3+37*x+4*x^4-4-43*x^2),            (-5*x-3-
5*x^2+x^3)/(-4*x^3+37*x+4*x^4-4-43*x^2)]
[    -(2*x^3-2*x^2+5*x-10)/(-4*x^3+37*x+4*x^4-
4-43*x^2),                    -2*(x^3+2*x^2-x-2)/(-
4*x^3+37*x+4*x^4-4-43*x^2),            (2*x^3-
3*x-1)/(-4*x^3+37*x+4*x^4-4-43*x^2)]
[            (7*x^2-24*x-8)/(-4*x^3+37*x+4*x^4-
4-43*x^2),                    (x^2-7*x-4)/(-
4*x^3+37*x+4*x^4-4-43*x^2),                    -
x*(3*x-13)/(-4*x^3+37*x+4*x^4-4-43*x^2)]
```

The above expression for the inverse matrix in symbolic form is very complicated. There are techniques in MATLAB and the Symbolic Math Toolbox that can be used to simplify the above expression but will not be used here. The reason is that the above expression can be simplified by hand by factoring out the determinant that appears in the denominators of the elements of the matrix.

In the next chapter, we will discuss programming in MATLAB concentrating on using scripts and functions.

Exercises

Solve all the exercises using MATLAB. All the needed MATLAB commands for these exercises were presented in this chapter. Note that Exercises 49-54 require the use of the MATLAB Symbolic Math Toolbox.

1. Generate the following rectangular matrix in MATLAB. What is the size of this matrix?

$$A = \begin{bmatrix} 3 & 0 & -2 \\ 1 & 3 & 5 \end{bmatrix}$$

2. In Exercise 1 above, extract the element in the second row and second column of the matrix A.

3. In Exercise 1 above, generate a new matrix B of the same size by multiplying the matrix A by the number $\dfrac{3\pi}{2}$.

4. In Exercise 3 above, extract the element in the first row and third column of the matrix B.

5. In Exercise 3 above, extract the sub-matrix of the elements in common between the first and second rows and the second and third columns of the matrix B. What is the size of this new sub-matrix?

6. In Exercise 3 above, determine the size of the matrix B using the MATLAB command `size`?

7. In Exercise 3 above, determine the largest of the number of rows and columns of the matrix B using the MATLAB command `length`?

8. In Exercise 3 above, determine the number of elements in the matrix B using the MATLAB command `numel`?

9. In Exercise 3 above, determine the total sum of each column of the matrix B? Determine also the minimum value and the maximum value of each column of the matrix B?

10. Combine the three vectors [1 3 0 -4] , [5 3 1 0], and [2 2 -1 1] to obtain a new matrix of size 3x4.

11. Perform the operations of matrix addition and matrix subtraction on the following two matrices:

$$R = \begin{bmatrix} 1 & 2 & 0 \\ 7 & 5 & -3 \\ 3 & 1 & 1 \end{bmatrix} \quad , \quad S = \begin{bmatrix} 1 & 3 & -2 \\ 3 & 5 & 7 \\ 2 & 3 & 0 \end{bmatrix}$$

12. In Exercise 11 above, multiply the two matrices R and S element-by-element.

13. In Exercise 11 above, divide the two matrices R and S element-by-element.

14. In Exercise 11 above, perform the operation of matrix multiplication on the two matrices R and S. Do you get an error? Why?

15. Add the number 5 to each element of the matrix X given below:

$$X = \begin{bmatrix} 1 & -2 & 0 & 1 \\ 2 & 3 & 6 & 2 \\ -3 & 5 & 2 & 1 \\ 5 & -2 & 4 & 4 \end{bmatrix}$$

16. In Exercise 15 above, subtract the number 3 from each element of the matrix X.

17. In Exercise 15 above, multiply each element of the matrix X by the number -3.

18. In Exercise 15 above, divide each element of the matrix X by the number 2.

19. In Exercise 15 above, perform the following multiple scalar operation $-3 * X / 2.4 + 5.5$.

20. Determine the sine, cosine, and tangent of the matrix B given below (element-by-element).

$$B = \begin{bmatrix} \dfrac{\pi}{3} & \dfrac{2\pi}{3} \\ \dfrac{2\pi}{3} & \pi \end{bmatrix}$$

21. In Exercise 20 above, determine the square root of the matrix B element-by-element.

22. In Exercise 20 above, determine the true square root of the matrix B.

23. In Exercise 20 above, determine the exponential of the matrix B element-by-element.

24. In Exercise 20 above, determine the true exponential of the matrix B.

25. In Exercise 20 above, determine the natural logarithm of the matrix B element-by-element.

26. In Exercise 20 above, determine the true natural logarithm of the matrix B.

27. In Exercise 20 above, perform the exponential operation 4^B.

28. In Exercise 27 above, repeat the same exponential operation but this time element-by-element.

29. In Exercise 20 above, perform the operation B^4.

30. Generate a rectangular matrix of 1's of size 2x3.

31. Generate a rectangular matrix of 0's of size 2x3.

32. Generate a rectangular identity matrix of size 2x3.

33. Generate a square matrix of 1's of size 4.

34. Generate a square matrix of 0's of size 4.

35. Generate a square identity matrix of size 4.

36. Determine the transpose of the following matrix:

$$C = \begin{bmatrix} 1 & 2 & -3 & 0 \\ 2 & 5 & 2 & -3 \\ 1 & 3 & 7 & -2 \\ 2 & 3 & -1 & 3 \end{bmatrix}$$

37. In Exercise 36 above, perform the operation $C + C'$. Do you get a symmetric matrix?

38. In Exercise 36 above, extract the diagonal of the matrix C.

39. In Exercise 36 above, extract the upper triangular part and the lower triangular part of the matrix C.

40. In Exercise 36 above, determine the determinant and trace of the matrix C. Do you get scalars?

41. In Exercise 36 above, determine the inverse of the matrix C.

42. In Exercise 41 above, multiply the matrix C by its inverse matrix. Do you get the identity matrix?

43. In Exercise 36 above, determine the norm of the matrix C.

44. In Exercise 36 above, determine the eigenvalues of the matrix C.

45. In Exercise 36 above, determine the coefficients of the characteristic polynomial of the matrix C.

46. In Exercise 36 above, determine the rank of the matrix C.

47. Generate a square random matrix of size 5.

48. Generate a square magic matrix of size 7. What is the sum of each row, column or diagonal in this matrix.

49. Generate the following two symbolic matrices:

$$ X = \begin{bmatrix} 1 & x \\ x-2 & x^2 \end{bmatrix} \quad , \quad Y = \begin{bmatrix} \dfrac{x}{2} & 3x \\ 1-x & 4 \end{bmatrix} $$

50. In Exercise 49 above, perform the matrix subtraction operation $X - Y$ to obtain the new matrix Z.

51. In Exercise 50 above, determine the transpose of the matrix Z.

52. In Exercise 50 above, determine the trace of the matrix Z.

53. In Exercise 50 above, determine the determinant of the matrix Z.

54. In Exercise 50 above, determine the inverse of the matrix Z.

8. Programming

In this chapter we introduce the basics of programming in MATLAB. The material presented in this chapter is not intended to be comprehensive or exhaustive but merely an introduction[23] to programming in MATLAB. In MATLAB programming, the list of commands and instructions are usually stored in a text file called an M-file (short for MATLAB file). These M-files can be of two kinds – either script files or function files. These files can be created or opened from the File menu by clicking on New or Open, then clicking on M-File. These files typically have the .m extension that is associated with MATLAB.

We will discuss script files first. The examples presented are simple. The reader can write more complicated examples based on the material presented here. Script files are used to store MATLAB scripts. A script is defined in MATLAB as a sequence of MATLAB commands. A script can span several lines. For example, here is a MATLAB script:

```
% This is an example
cost = 50
profit = 10
sale_price = cost + profit
```

Note that the first line in the script above is a comment line – this is optional. Store the above example in an M-file called example1.m then run the example by typing example1 at the command line to get:

```
>> example1

cost =
```

---

[23] The interested reader may consult the references listed at the end of this book for more details about programming in MATLAB.

```
    50


profit =

    10


sale_price =

    60
```

There are several remarks about the previous example. The first remark is that a script needs to be defined first (in an M-file) then executed by typing the name of the file (without the .m extension) on the command line. Note the use of the comment symbol % on the first line which is a comment line. The use of comments in scripts is optional but is useful to the reader. Note also that sometimes we do not need the results of all the commands displayed by MATLAB. For this purpose, we can use semicolons to suppress the outputs that are not needed. For example, here is the same example above with the output suppressed with semicolons except the last line:

```
% This is an example
cost = 50;
profit = 10;
sale_price = cost + profit
```

Store the above script in a new M-file and call it example2.m. Executing the above example by typing example2 at the command line will result in the following output:

```
>> example2

sale_price =
```

60

The second type of M-files in MATLAB is called a function file. These are more structured[24] than script files. In general, scripts are easier to write than functions but functions have their advantages. In fact, many of the commands used by MATLAB are stored in function files. Functions usually contain a sequence of MATLAB commands possibly spanning several lines. In addition, a function has a function name[25] and one or more arguments. In addition, a function returns a value[26]. In contrast, a script does not have any arguments. For example, here is a function called `area(r)` that is stored in an M-File called `area.m`

```
function area(r)
% This is a function
area = pi*r^2
```

The above function calculates the area of a circle of radius r. The first line of the function file must have the word `function` followed by the name of the function and the argument(s) in parentheses. The last line of the function usually includes a calculation of the variable that is used for the name of the function. Now, run the above function by typing the name of the function followed by the argument in parentheses along with its value on the command line. Here is the result:

```
>> area(3)

area =
```

---

[24] This means that there are rules for writing functions in MATLAB.

[25] The name of the function must be exactly the same as the name of the function file it is stored in.

[26] Consult a book on programming for more details about functions and return values. In other programming languages, functions are called procedures or subroutines and have a different structure than in MATLAB.

```
28.2743
```

A function may be executed several times with a different value for the argument each time. For example here is another execution of the previous function with the value 5 replacing 3^{27}:

```
>> area(5)

area =

    78.5398
```

A function may be executed as many times as needed. However, it needs to be defined only once in an M-file. Here is an example of a function that calculates the perimeter of a rectangle with sides a and b:

```
function perimeter(a,b)
% This is another example of a function
perimeter = 2*(a+b)
```

It is clear that the above function has two arguments, namely a and b. Now, execute[28] the above function three times, each time for a certain rectangle with a specified length and width as follows:

```
>> perimeter(2,3)

perimeter =

    10

>> perimeter(3,7)
```

---

[27] No units are used in this example.

[28] The two words "execute" and "run" are used interchangeably in this chapter. They are basically equivalent for our purpose in this chapter.

```
perimeter =

    20

>> perimeter(1.2,5.3)

perimeter =

    13
```

It is clear that the first execution above calculates the perimeter for a rectangle with sides 2 and 3, the second execution calculates the perimeter with sides 3 and 7, while the third execution calculates the perimeter with sides 1.2 and 5.3[29].

Next, we will discuss briefly several constructs[30] that are used in programming in MATLAB - inside scripts and functions. In particular, we will discuss loops (the For loop and the While loop) and decisions (the If Elseif construct and the Switch Case construct).

In MATLAB loops, several command lines are executed repeatedly over many cycles depending on certain parameters in the loop. There are two types of loops in programming in MATLAB – the For loop and the While loop. Consider the following script file which is to be stored under the file name example3.m – this is an example of a For loop.

```
% This is an example of a FOR loop
for n = 1:10
    x(n) = n^2;
end
```

---

[29] Note that no units are specified for the dimensions used in this example.
[30] It should be noted that these programming constructs are also available in other programming languages but they have a different structure, i.e. different rules for their implementation.

Now, run the above script by typing the command `example3` at the command prompt as follows:

`>> example3`

No output will be displayed because of the presence of the semicolon in the script file above. However, to find the value of the variable x, you need only to type x at the command prompt as follows:

`> x`

`x =`

```
    1      4      9     16     25     36     49
   64     81    100
```

It is seen from the above example that the For loop cycles over the values of n from 1 to 10 – exactly 10 times. In each cycle, the value of the vector x is calculated using the quadratic formula inside the For loop. The final result for the vector x is displayed above – it is seen that the vector x has exactly 10 elements.

Here is another example of a For loop to be stored in a script file called `example4.m`

```
%This is an example of another FOR loop
for n = 1:3
    for m = 1:3
        y(m,n) = m^2 + m*n + n^2;
    end
end
```

Actually the above example contains two For loops that are nested. Now, run the above script file by typing `example4` at the command prompt as follows, then type y to get the values of the elements of the matrix y:

```
>> example4
>> y

y =

     3      7     13
     7     12     19
    13     19     27
```

It is seen that two nested For loops in this example can produce a matrix with a single assignment inside the For loops.

Another way to use loops in MATLAB is to use the While loop instead of the For loop. Here is an example of a While loop to be stored in a script file called example5.m

```
% This is an example of a While loop
tol = 0.0;
n = 0;
while tol < 10
      n = n + 1;
      tol = tol + 2;
end
```

Now, run the above example by typing example5 at the command prompt, then type tol and n to get their values after executing the script:

```
>> example5
>> n

n =
```

```
    5
```

```
>> tol
```

```
tol =
```

```
    10
```

It is seen from the above example that the loop will continue to cycle as long as the value of the variable `tol` is less than[31] 10. As soon as the value of `tol` becomes equal to or larger than 10, the loop ends. The final values of the variables are then displayed by MATLAB with the appropriate commands. Note also the use of the semicolons in the script file.

Next, we discuss implementing decisions in programming in MATLAB. There are several constructs in MATLAB to use for decisions like the `If` construct, the `If Else Then` construct, the `If Elseif Then` construct, and the `Switch Case` construct. We will discuss these constructs briefly in the remaining part of this chapter.

Consider the following script file containing an `If` construct to be stored in a script file called `example6.m`

```
% This is an example using If Then
boxes = 10;
cost = boxes*3;
if boxes > 7
    cost = boxes*2.5;
end
```

---

[31] The logical operator "less than" is represented in MATLAB by the symbol <. We do not cover logical operators in this book but this is an instance where their use is needed.

Now, run the above example by typing `example6` at the command line. Then type the variable `cost` to see which answer we get as follows:

```
>> example6
>> cost

cost =

    25
```

It is clear from the above example that we get 25 as the value of the variable `cost`. This is because the statement inside the `If` construct was executed. Actually, what happens is as follows: First, MATLAB computes the value of the variable `cost` as 30 using the first assignment statement for the variable `cost`. However, once we enter the `If` construct, the value of the variable `cost` may change depending on certain parameters. In this case, the value will change only if the value of the variable `boxes` is greater than 7. In our case, the value of the variable `boxes` is 10, so the effect of the `If` construct is to change the computation of the value of the variable `cost` from 30 to 25 using the second assignment statement for the variable `cost` – the one inside the `If` construct.

Here is the same example above but written as a function in a function file called `cost.m`. The function is called `cost(boxes)` and has one argument only. Writing the example as a function and not as a script will give us the option of changing the value of the variable `boxes` at the command prompt to see what effect it will have on the value of the variable `cost`. Here is the definition of the function:

```
function cost(boxes)
% This is an example of an If construct
cost = boxes*3
if boxes > 7
```

```
          cost = boxes*2.5
     end
```

Now, execute the above function using different values for the variable boxes. In this case, we will execute it three times to see the effect it has on the value of the variable cost as follows:

```
>> cost(4)

cost =

    12

>> cost(6)

cost =

    18

>> cost(8)

cost =

    24


cost =

    20
```

It is seen that the values of the variable cost for the first two executions were obtained directly using the first assignment statement for cost – this is because the value of the variable boxes is less than or equal to 7 in these two cases. However, for the third execution, the value of cost was calculated initially to be

24 but changed to 20 finally and correctly because the value of the variable boxes is larger than 7 in this last case.

Next, we discuss the If Elseif[32] construct using a slightly modified version of the function cost(boxes). In this case, we call the function cost2(boxes) and define it as follows to be stored in a function file called cost2.m

```
function cost2(boxes)
% This is an example of an If Elseif construct
if boxes < 7
    cost = boxes*2.5
elseif boxes < 9
    cost = boxes*2
elseif boxes > 9
    cost = boxes*1.5
end
```

Now, we execute the above function several times to test the results we get. We will execute it three times with the values of 6, 8, and 10 for the variable boxes. Here is what we get:

```
>> cost2(6)

cost =

    15

>> cost2(8)

cost =
```

---

[32] Note that there other variations of the above constructs used for implementing decisions in MATLAB, e.g. the If Else construct (see the last section in this chapter on the Symbolic Math Toolbox to see an example of using this construct). Consult a book dedicated on programming in MATLAB for more details about these constructs.

```
    16
```

```
>> cost2(10)
```

```
cost =
```

```
    15
```

In the above example, it is seen that the appropriate branch of the If Elseif construct and the appropriate assignment statement is used for the computation of the value of the variable cost depending on the value of the variable boxes. Note that several Elseif 's were needed to accomplish this. A more efficient way of doing this is to use the Switch Case construct. This is achieved by defining a new function cost3(boxes) to be stored in the function file cost3.m as follows (note that the functions cost2 and cost3 are not exactly equivalent):

```
function cost3(boxes)
% This is an example of the Switch Case construct
switch boxes
    case 6
        cost = boxes* 2.5
    case 8
        cost = boxes* 2
    case 10
        cost = boxes*1.5
    otherwise
        cost = 0
end
```

Now, execute the above function three times as before with the three values of 6, 8, and 10 for the variable boxes. This is what we get (exactly as before):

```
>> cost3(6)
```

```
cost =

    15

>> cost3(8)

cost =

    16

>> cost3(10)

cost =

    15
```

Programming with the MATLAB Symbolic Math Toolbox

In this section we discuss scripts and functions using the MATLAB Symbolic Math Toolbox. Consider the following script where we square a symbolic matrix:

```
% This is an example with the Symbolic Math Toolbox
syms x
A = [x 1-x x+2 ; 2*x -3*x x ; 2-x 1-x x*2]
B = A^2
```

In the above script example, we define a symbolic variable x, then define the symbolic matrix A. Finally, we obtain the symbolic matrix B as the square of A. Now, store the above script in a script file called example7.m then run the script by typing example7 at the command prompt. This is what we get:

```
>> example7
```

A =

```
[    x,    1-x,    x+2]
[  2*x,   -3*x,      x]
[  2-x,    1-x,    2*x]
```

B =

```
[   x^2+2*(1-x)*x+(x+2)*(2-x),              -2*(1-
x)*x+(x+2)*(1-x),              3*x*(x+2)+(1-x)*x]
[              -4*x^2+x*(2-x),                  3*(1-
x)*x+9*x^2,                  2*x*(x+2)-x^2]
[          3*x*(2-x)+2*(1-x)*x,              (2-x)*(1-
x)-(1-x)*x,   (x+2)*(2-x)+(1-x)*x+4*x^2]
```

Our next example will use a function file. Consider the following function SqureRoot(matrix) to be stored in a function file called SquareRoot.m

```
function SquareRoot(matrix)
% This is a function with Symbolic Variables
y = det(matrix)
z = subs(y,1)
if z < 1
    M = 2*sqrt(matrix)
else
    M = sqrt(matrix)
end
```

In the above function, the If Else construct is used to check the determinant of the matrix that is passed to the function. Once the determinant of the matrix is computed symbolically, we use the subs command to substitute the value of 1 for the symbolic variable x. If the value of the determinant of the matrix is less than 1, then we take twice the square root of the matrix (element-by element)[33], otherwise, we take the square root of the matrix (element-by-element). Now, execute the above function with the following symbolic matrix to see the outcome:

```
>> syms x
>> M = [ 1-x x^2 ; x+3 x]

M =

[ 1-x,  x^2]
[ x+3,    x]

>> SquareRoot(M)

y =

x-4*x^2-x^3

  z =

    -4

M =

[ 2*(1-x)^(1/2),  2*(x^2)^(1/2)]
[ 2*(x+3)^(1/2),      2*x^(1/2)]
```

In the next chapter, we will discuss graphs and plotting in MATLAB.

---

[33] The true square root of a matrix using the sqrtm command is not defined for symbolic variables in MATLAB.

Exercises

Solve all the exercises using MATLAB. All the needed MATLAB commands for these exercises were presented in this chapter. Note that Exercises 10-11 require the use of the MATLAB Symbolic Math Toolbox.

1. Write a script of four lines as follows: the first line should be a comment line, the second and third lines should have the assignments `cost = 200` and `sale_price = 250`, respectively. The fourth line should have the calculation `profit = sale_price - cost`. Store the script in a script file called `example8.m`. Finally run the script file.

2. Write a function of three lines to calculate the volume of a sphere of radius `r`. The first line should include the name of the function which is `volume(r)`. The second line should be a comment line. The third line should include the calculation of the volume of the sphere which is $\frac{4}{3}\pi r^3$.

    Store the function in a function file called `volume.m` then run the function with the value of `r` equal to 2 (no units are used in this exercise).

3. Write a function with two arguments to calculate the area of a rectangle with sides `a` and `b`. The function should have three lines. The first line should include the name of the function which is `RectangleArea(a,b)`. The second line should be a comment line. The third line should include the calculation of the area of the rectangle with is the product `a*b`. Store the function in a function file called `RectangleArea.m` then run the function twice as follow: the first execution with the values 3 and 6, while the second execution with the values `2.5` and `5.5`.

4. Write a script containing a For loop to compute the vector x to have the values $x(n) = n^3$ where n has the range from 1 to 7. Include a comment line at the beginning. Store the script in a script file called example9.m then run the script and display the values of the elements of the vector x.

5. Write a script containing two nested For loops to compute the matrix y to have the values $y(m,n) = m^2 - n^2$ where both m and n each has the range from 1 to 4. Include a comment line at the beginning. Store the script in a script file called example10.m then run the script and display the values of the elements of the matrix y.

6. Write a script containing a While loop using the two variables tol and n. Before entering the While loop, initialize the two variables using the assignments tol = 0.0 and n = 3. Then use the two computations n = n + 1 and tol = tol + 0.1 inside the While loop. Make the loop end when the value of tol becomes equal or larger than 1.5. Include a comment line at the beginning. Store the script in a script file called example11.m then run the script and display the values of the two variables tol and n.

7. Write a function called price(items) containing an If construct as follows. Let the price of the items be determined by the computation price = items*130 unless the value of the variable items is greater than 5 – then in this case the computation price = items*160 should be used instead. Include a comment line at the beginning. Store the function in a function file called price.m then run the function twice with the values of 3 and 9 for the variable items. Make sure that the function displays the results for the variable price.

8. Write a function called price2 (items) containing an If Elseif construct as follows. If the value of the variable items is less than 3, then compute the variable price2 by multiplying items by 130. In the second case, if the value of the variable items is less than 5, then compute the variable price2 by multiplying items by 160. In the last case, if the value of the variable items is larger than 5, then compute the variable price2 by multiplying the items by 200. Include a comment line at the beginning. Store the function in a function file called price2.m then run the function three times – with the values of 2, 4, and 6. Make sure that the function displays the results for the variable price2.

9. Write a function called price3 (items) containing a Switch Case construct. The function should produce the same results obtained in Exercise 8 above. Include a comment line at the beginning. Store the function in a function file called price3.m then run the function three times – with the values of 2, 4, and 6. Make sure that the function displays the results for the variable price3.

10. Write a script file to store the following symbolic matrix A then calculate its third power $B = A^3$. Include a comment line at the beginning. Store the script in a script file called example12.m then run the script to display the two matrices A and B.

$$A = \begin{bmatrix} \dfrac{x}{2} & 1-x \\ x & 3x \end{bmatrix}$$

11. Write a function called SquareRoot2 (matrix) similar to the function SquareRoot (matrix) described at the end of this chapter but with the following change. Substitute

the value of 1.5 instead of 1 for the symbolic variable x. Make sure that you include a comment line at the beginning. Store the function in a function file called SquareRoot2.m then run the function using the following symbolic matrix:

$$M = \begin{bmatrix} 2 & x & 0 \\ 3-x & 5 & -x \\ x+2 & 1 & 3 \end{bmatrix}$$

9. Graphs

In this chapter we explain how to plot graphs in MATLAB. Both two-dimensional and three-dimensional graphs are presented. A graph in MATLAB appears in its own window (not on the command line). First, we will consider two-dimensional or planar graphs. To plot a two-dimensional graph, we need two vectors. Here is a simple example using two vectors x and y along with the MATLAB command plot:

```
>> x = [1 2 3 4 5]

x =

        1        2        3        4        5

>> y = [3 9 12 10 6]

y =

        3        9       12       10        6

>> plot(x,y)
```

The resulting graph is displayed in its own window and is shown in Figure 9.1. Note how the command plot was used above along with the two vectors x and y. This is the simplest use of this command in MATLAB.

We can add some information to the above graph using other MATLAB commands that are associated with the plot command. For example, we can use the title command to add a title to the graph. Also, we can use the MATLAB commands xlabel and ylabel to add labels to the x-axis and the y-axis. But before using these three commands (title, xlabel, ylabel), we need to keep the plotted graph in its window. For this we use the `hold on` command as follows:

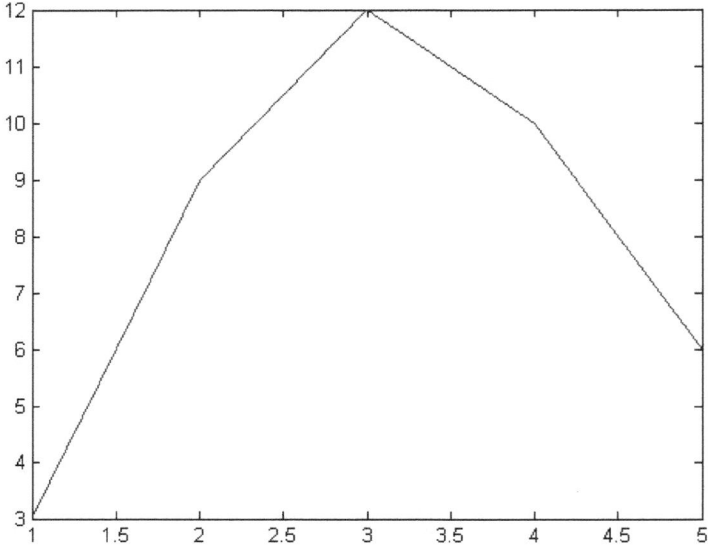

Figure 9.1: A simple use of the plot command

```
>> hold on
>> title('This is an example of a simple graph')
>> xlabel('x')
>> ylabel('y')
```

After executing the above commands, the plotted graph appears as shown in Figure 9.2 with the title and axes labels clearly displayed.

Let us now plot the mathematical function $y = x^2 - 4$ in the range x from -3 to $+3$. First, we need to define the vector x, then calculate the vector y using the above formula. Then we use the `plot` command as usual for the two vectors x and y. Finally, we use the other commands to display the title and axis label information on the graph.

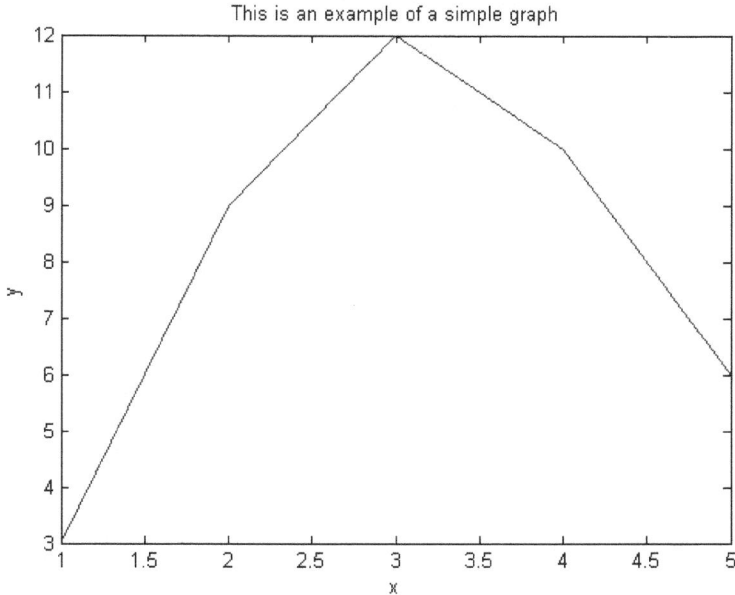

This is an example of a simple graph

Figure 9.2: Title and axis information displayed on graph

The following are the needed commands. The resulting graph is shown in Figure 9.3.

```
>> x = [-3 -2 -1 0 1 2 3]

x =

      -3      -2      -1       0       1       2       3

>> y = x.^2 -3

y =

       6       1      -2      -3      -2       1       6
```

```
>> plot(x,y)
>> hold on
>> title('This is another example of the plot command')
>> xlabel('x-axis')
>> ylabel('y-axis')
```

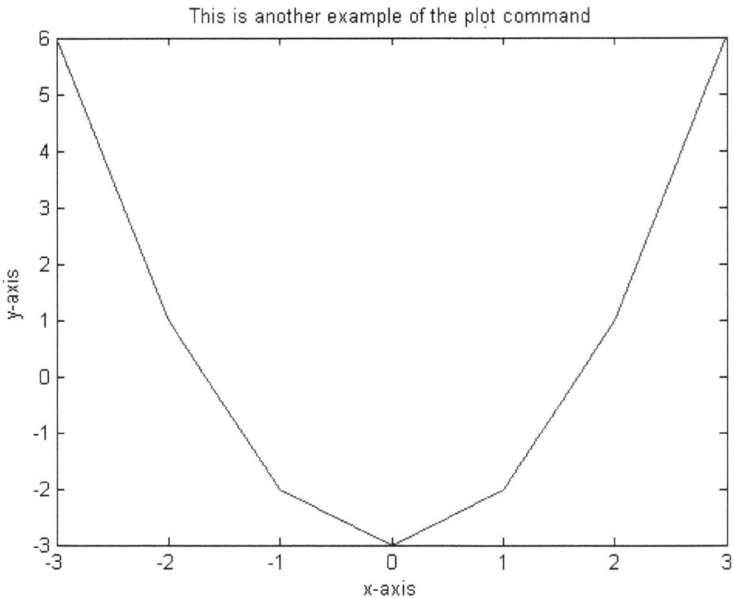

Figure 9.3: Example of plotting a mathematical function

It is seen from the above graphs that they are plotted with solid lines. The line style may be changed from solid to dotted, dashed, or dash-dot lines by changing the parameters in the plot command. Use the command plot(x,y,':') to get a dotted line, the command plot(x,y,'--') to get a dashed line, the command plot(x,y,'-.') to get a dash-dot line, and the default command plot(x,y,'-') to get a solid line.

In addition, the color of the plotted line may be changed. Use the command plot(x,y,'b') to get a blue line. Replace the b with g for green, r for red, c for cyan, m for magnetta, y for yellow, k for black, and w for white. Consult the MATLAB manual for a full list of the color options. Note that colors are not displayed in this book.

In addition to changing the line style and color, we can include symbols at the plotted points. For example, we can use the circle or cross symbols to denote the plotted points. Use the command plot(x,y,'o') to get a circle symbol. Replace the o with x to get a cross symbol, + for a plus sign symbol, * for a star symbol, s for a square symbol, d for a diamond symbol, and . for a point symbol. There are other commands for other symbols like different variations of triangles, etc. Again, consult the MATLAB manual for the full list of available symbols.

The above options for line style, color, and symbols can be combined in the same command. For example use the command plot(x,y,'rx:') to get a red dotted line with cross symbols. Here is the previous example with a red dotted line with cross symbols but without the title and label information (note that no colors appear in this book). See Figure 9.4 for the resulting graph.

```
>> plot(x,y,'rx:')
```

Next, consider the following example where we show another use of the plot command. In this example, we plot two curves on the same graph. We use the modified plot command as plot(x,y,x,z) to plot the two curves. MATLAB plots the two mathematical functions y and z (on the y-axis) as a function of x (on the x-axis). We also use the MATLAB command grid to show a grid on the plot. Here are the needed commands[34]. The resulting graph is shown in Figure 9.5.

```
>> x = 0:pi/20:2*pi;
```

------

[34] These commands may be entered individually on the command prompt or stored as a script in a script file. See Chapter 8 for details about scripts and script files.

```
>> y = sin(x);
>> z = cos(x);
>> plot(x,y,'-',x,z,'--')
>> hold on;
>> grid on;
>> title('Sine and Cosine Functions')
>> xlabel('x-axis')
>> ylabel('sine and cosine')
>> legend('sinx','cosx')
```

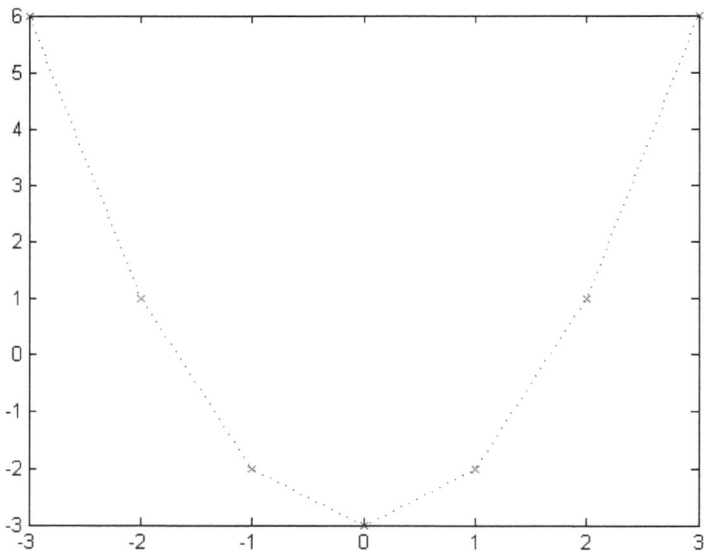

Figure 9.4: A red dotted line with cross symbols

Note in the above example that we used semicolons to suppress the output. Also, note the use of the MATLAB command legend to get a legend displayed at the top right corner of the graph. We can further use several axis commands to customize the appearance of the two axes including the tick marks, but this is beyond the scope of this book. Consult the MATLAB manual for further details about these commands.

There are some interactive plotting tools available in MATLAB to further customize the resulting graph. These tools are available from the menus of the graph window – the window where the graph appears.

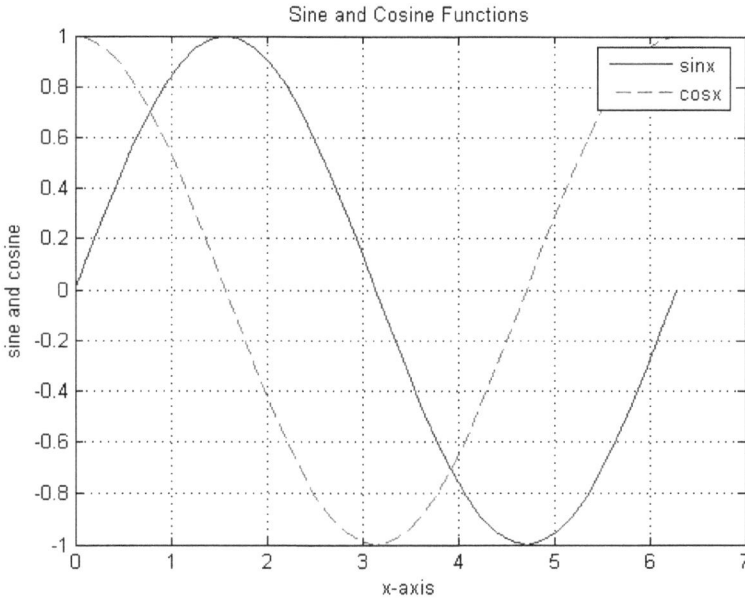

Figure 9.5: Example of two curves on the same graph

There are other specialized plots[35] that can be drawn automatically by MATLAB. For example, there are numerous plots that are used in statistics that can be generated by MATLAB. In particular, a pie chart can be generated by using the MATLAB command pie, a histogram can be generated by using the MATLAB command hist, and a rose diagram can be generated by using the

---

[35] Several logarithmic plots can also be generated by using the MATLAB command loglog. Semi-logarithmic plots can also be generated by using the MATLAB commands semilogx or semilogy but their use is beyond the scope of this book.

MATLAB command `rose`. In addition, graphs of curves with polar coordinates can be generated by using the MATLAB command `polar`.

Our final two-dimensional graph will feature the use of the MATLAB command `subplot`. The use of this command will enable you to present several plot diagrams on the same graph. In this example, we use four instances of the `subplot` command to show four diagrams on the same plot. We will plot four mathematical functions – each on its own diagram – but the four diagrams will appear on the same graph. We can do this by playing with the parameters of the `subplot` command. Note that the diagrams are shown without title or axis labels in order to emphasize the use of the `subplot` command. Here are the necessary MATLAB commands[36] used with semicolons to suppress the output:

```
>> x = [ 1 2 3 4 5 6 7 8 9 10];
>> y = x.^2;
>> z = sqrt(x);
>> w = 2*x - 3;
>> v = x.^3;
>> subplot(2,2,1)
>> hold on;
>> plot(x,y)
>> subplot(2,2,2)
>> plot(x,z)
>> subplot(2,2,3)
>> plot(x,w)
>> subplot(2,2,4)
>> plot(x,v)
```

Note the use of the three parameters of the `subplot` command. The first two parameters (2,2) indicate that the graph area should be divided into four quadrants with each row and column

---

[36] These commands may be entered individually on the command prompt or stored in as a script in a script file. See Chapter 8 for details about scripts and script files.

comprising of two sub-areas for plots. The third parameter indicates in which sub-area the next plot will appear. Note also that each `subplot` command is followed by a corresponding `plot` command. Actually, each `subplot` command reserves a specific area for the plot while the subsequent corresponding `plot` command does the actual plotting. The resulting graph is shown in Figure 9.6. You can control the number and arrangement of the diagrams that are displayed by controlling the three parameters of the `subplot` command.

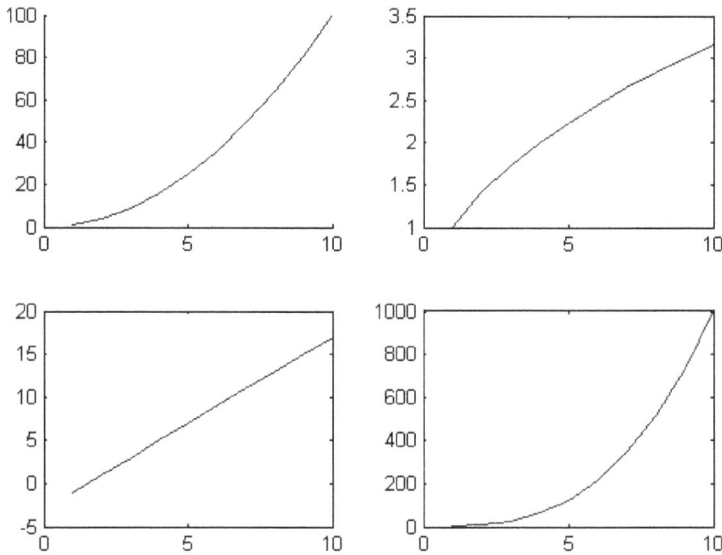

Figure 9.6: Use of the MATLAB command subplot

Next, we will discuss how to generate three-dimensional graphs with MATLAB. There are mainly four commands that can be used in MATLAB for plotting three-dimensional graphs. The first command is `plot3` which is used to plot curves in space. It requires three vectors as input. Here is an example of this command followed by the resulting graph as shown in Figure 9.7.

```
>> x = [1 2 3 4 5 6 7 8 9 10]

x =

     1      2      3      4      5      6      7
8      9     10

>> y = [1 2 3 4 5 6 7 8 9 10]

y =

     1      2      3      4      5      6      7
8      9     10

>> z = sin(x).*cos(y)

z =

  Columns 1 through 6

     0.4546    -0.3784    -0.1397     0.4947    -
0.2720    -0.2683

  Columns 7 through 10

     0.4953    -0.1440    -0.3755     0.4565

>> plot3(x,y,z)
>> hold on;
>> grid on;
```

It should be noted that the title, xlabel, and ylabel commands can also be used with three-dimensional graphs. For the z-axis, the MATLAB command zlabel is also used in a way similar to the xlabel and ylabel commands.

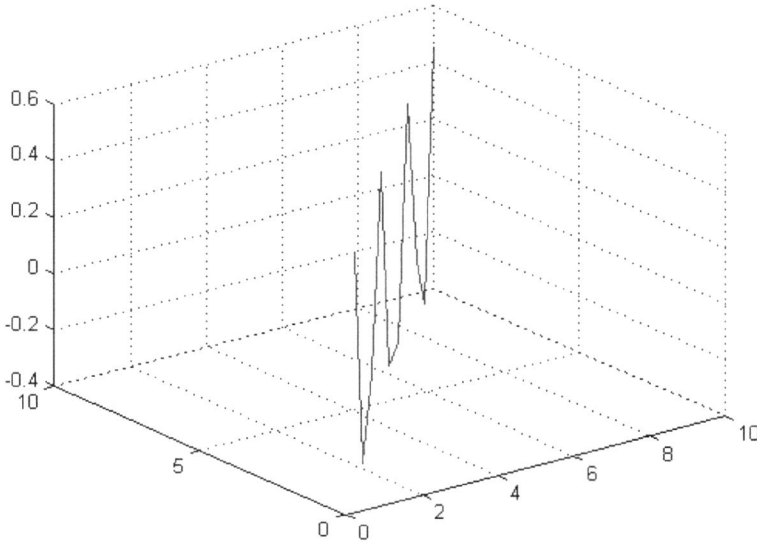

Figure 9.7: Example of using the plot3 command

The second command in MATLAB for three dimensional plots is mesh. The use of the mesh command is used to generate mesh surfaces. The simplest use of this command is with a matrix. It assumes that the matrix elements are values of a grid spanning the matrix dimensions. The values of the elements of the matrix are taken automatically along the z-axis while the numbers of rows and columns are taken along the x- and y-axes. Here is an example followed by the resulting graph as shown in Figure 9.8.

```
>> A = [10 14 20 37 5 17 11 ; 12 20 40 20 11
5 14 ; 30 51 12 17 20 30 2 ; 24 34 56 10 14
5 40 ; 34 12 33 12 26 10 15 ; 12 45 13 23 35
10 7 ; 10 20 13 34 32 10 7]

A =

      10     14     20     37      5     17     11
```

```
    12      20      40      20      11       5      14
    30      51      12      17      20      30       2
    24      34      56      10      14       5      40
    34      12      33      12      26      10      15
    12      45      13      23      35      10       7
    10      20      13      34      32      10       7
```

`>> mesh(A)`

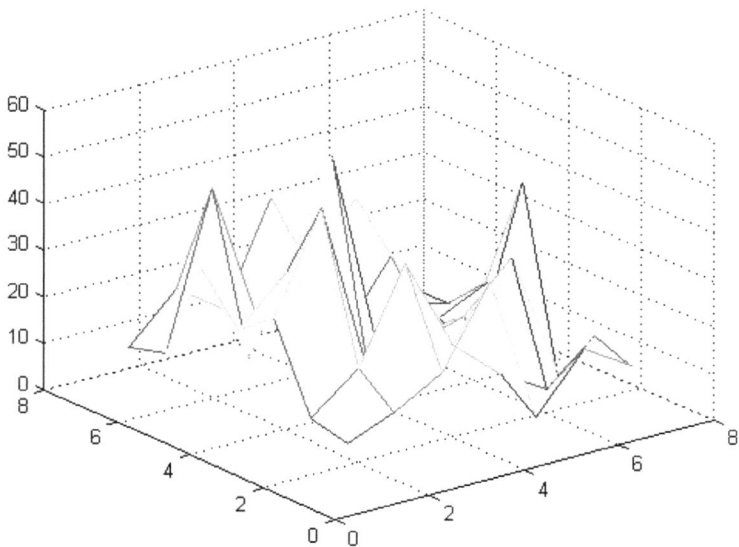

Figure 9.8: Use of the mesh command

If a matrix or a grid is not available, then a grid can be generated using the MATLAB command `meshgrid` but this is beyond the scope of this book. Again, the reader should use the title and axis information commands to display information on the graph but these are not shown in this example.

The third MATLAB command for three-dimensional plots is the `surf` command. Its use is similar to the `mesh` command. Here is an example of its use with the same matrix A of the previous example followed by the resulting graph as shown in Figure 9.9.

```
>> surf(A)
```

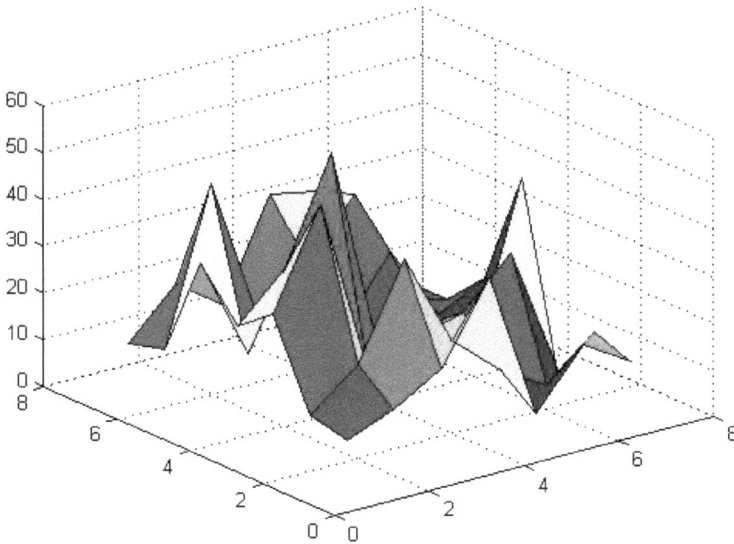

Figure 9.9: Use of the surf command

The above three-dimensional plot is usually shown in color in MATLAB but we are not able to display colors in this book. Again, the MATLAB command meshgrid may be used to generate a grid for the three-dimensional plot if one is not available. Again, the reader should use the title and axis information commands to display these types of information on the plot above but these are not used in this example.

The fourth command for three-dimensional plots is the MATLAB command contour. The use of this command produces contour plots of three-dimensional surfaces. The use of this command is similar to the mesh and surf commands. Here is an example of its use with the same matrix A that was used previously followed by the resulting graph as shown in Figure 9.10.

```
>> contour(A)
```

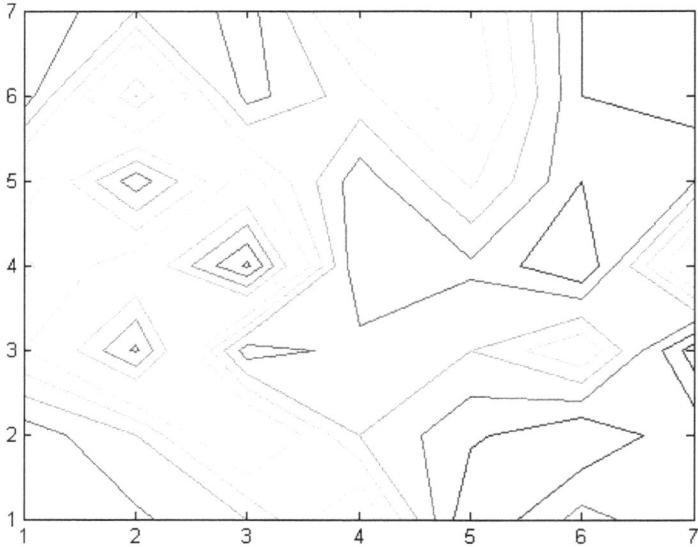

Figure 9.10: Use of the contour command

The above contour plot is usually shown in color in MATLAB but we are not able to display colors in this book. Again, the MATLAB command meshgrid may be used to generate a grid for the contour plot if one is not available. The reader can add the contour heights to the graph using the MATLAB command clabel.

There are some variations of the mesh and surf commands in MATLAB. For example, the meshc and surfc commands produce the same mesh and surface plots as the commands mesh and surf but with the contour plot appearing underneath the surface. Here is an example of the use of the surfc command with the same matrix A used previously followed by the resulting graph as shown in Figure 9.11.

```
>> surfc(A)
```

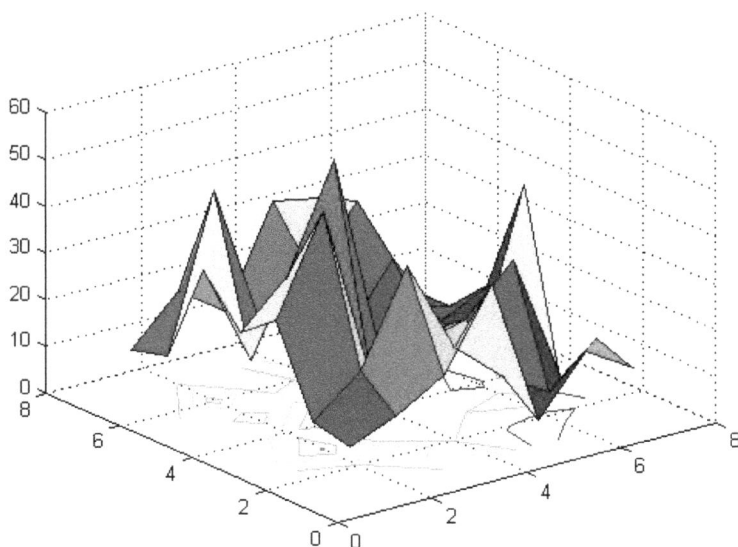

Figure 9.11: Use of the surfc command

Another variation is the meshz command which produces the same three-dimensional plot as the mesh command but with a zero plane underneath the surface. The use of this command is very simple and is not shown here.

In the next chapter, we discuss solution of equations in MATLAB.

Exercises

Solve all the exercises using MATLAB. All the needed MATLAB commands for these exercises were presented in this chapter.

1. Plot a two-dimensional graph of the two vectors x = [1 2 3 4 5 6 7] and y = [10 15 23 43 30 10 12] using the plot command.

2. In Exercise 1 above, add to the graph a suitable title along with labels for the x-axis and the y-axis.

3. Plot the mathematical function $y = 2x^3 + 5$ in the range x from -6 to $+6$. Include a title for the graph as well as labels for the two axes.

4. Repeat the plot of Exercise 3 above but show the curve with a blue dashed line with circle symbols at the plotted points.

5. Plot the two mathematical functions $y = 2\sin(\frac{x}{3})$ and $z = 2\cos(\frac{x}{3})$ such that the two curves appear on the same diagram. Make the range for x between 0 and 3π with increments of $\frac{\pi}{4}$. Distinguish the two curves by plotting one with a dashed line and the other one with a dotted line. Include the title and axis information on the graph as well as a grid and a legend.

6. Plot the following four mathematical functions each with its own diagram using the subplot command. The functions are $y = 2x^3 - 4$, $z = x + 1$, $w = 2 - \sqrt{x}$, and $v = x^2 + 3$. Use the vector x = [1 2 3 4 5 6 7 8 9 10] as the range for x. There is no need to show title or axis information.

7. Use the plot3 command to show a three-dimensional curve of the equation $z = 2\sin(xy)$. Use the vector x = [1 2 3 4 5 6

7 8 9 10] for both x and y. There is no need to show title or axis information.

8. Use the mesh command to plot a three-dimensional mesh surface of elements of the matrix M given below. There is no need to show title or axis information.

$$M = \begin{bmatrix} 0.1 & 0.2 & 0.5 & 0.7 & 0.3 \\ 0.2 & 0.8 & 0.5 & 0.6 & 0.3 \\ 0.5 & 0.9 & 0.9 & 0.4 & 0.4 \\ 0.4 & 0.4 & 0.5 & 0.7 & 0.9 \\ 0.1 & 0.3 & 0.4 & 0.6 & 0.8 \end{bmatrix}$$

9. Use the surf command to plot a three-dimensional surface of the elements of the matrix M given in Exercise 8 above. There is no need to show title or axis information.

10. Use the contour command to plot a contour map of the elements of the matrix M given in Exercise 8 above. There is no need to show the contour values on the graph.

11. Use the surfc command to plot a three-dimensional surface along with contours underneath it of the matrix M given in Exercise 8 above. There is no need to show title or axis information.

10. Solving Equations

In this chapter we discuss how to solve algebraic equations using MATLAB. We will discuss both linear and nonlinear algebraic equations. We will also discuss systems of simultaneous linear and nonlinear algebraic equations. First, let us solve the following simple linear equation for the variable x:

$$2x - 3 = 0$$

The solution of the above equation may be obvious to the reader but we will show how to solve it using MATLAB. The easiest method to do it in MATLAB would be to consider the left-hand-side of the equation as a polynomial of first degree and then find the roots of the polynomial using the MATLAB command roots. First, we write the above equation as a polynomial as follows:

$$p(x) = 2x - 3 = 0$$

We enter the coefficients of the above polynomial in a vector in MATLAB then use the roots command to find the root which is the solution to the above linear equation in x. Here are the needed commands:

```
>> p = [2 -3]

p =

    2    -3

>> roots(p)

ans =

    1.5000
```

It is clear that the correct solution for x is obtained which is 1.5. Next, let us show similarly how to find the solution of a quadratic equation in x by solving the following quadratic equation:

$$5x^2 + 3x - 4 = 0$$

We will consider the above equation as a polynomial of second degree in x. We will enter the coefficients of the polynomial in a vector in MATLAB followed by the `roots` command. Here are the needed MATLAB commands:

```
>> p = [5 3 -4]

p =

     5      3     -4

>> roots(p)

ans =

    -1.2434
     0.6434
```

It is clear from the above that two real solutions are obtained for x. Let us now solve a more complicated equation. Let us solve the following equation for x:

$$2x^5 - 3x^4 - 5x^3 + x^2 - 1 = 0$$

First, we consider the left-hand-side as a polynomial of fifth degree in x. Then we write the coefficients of this polynomial as a vector in MATLAB followed by the `roots` command as follows (note that we will enter 0 for the missing x in the above polynomial):

```
>> p = [2 -3 -5 1 0 -1]

p =

    2     -3     -5      1      0     -1

>> roots(p)

ans =

    2.4507
   -1.0000
   -0.6391
    0.3442 + 0.4480i
    0.3442 - 0.4480i
```

Note from the above that five values are obtained as the solution of the above equation. These values are the five roots of the polynomial given above. Note that three of these roots are real while two roots are complex conjugates of each other.

Next, let us discuss how to solve a system of linear simultaneous algebraic equations. In order to solve such systems there are several methods available with most of them involving the use of matrices and vectors. Consider the following simple system of two linear algebraic equations in the variables x and y:

$$x - 3y = 5$$
$$4x + 6y = 3$$

In order to solve the above system of equations, we need to re-write the system as a matrix equation. The coefficients of x and y on the left-hand-side of the equations are considered as the elements of a square matrix (of size 2x2 here) while the numbers on the right-hand-side are entered into a vector. The resulting equivalent matrix equation is written as follows:

$$\begin{bmatrix} 1 & -3 \\ 4 & 6 \end{bmatrix} \begin{Bmatrix} x \\ y \end{Bmatrix} = \begin{Bmatrix} 5 \\ 3 \end{Bmatrix}$$

The above matrix equation is in the form $[A]\{x\} = \{b\}$. The solution[37] of this equation is $\{x\} = [A]^{-1}\{b\}$. Thus we need to find the inverse of the coefficient matrix $[A]$ and multiply it with the constant vector $\{b\}$. Here are the needed MATLAB commands along with the solution:

```
>> A = [1 -3 ; 4 6]

A =

      1       -3
      4        6

>> b = [5 ; 3]

b =

      5
      3

>> x = inv(A)*b

x =

      2.1667
     -0.9444
```

Thus it is clear from the above that x = 2.1667 and y = -0.9444 are the solutions to the above system of equations. Note

---

[37] Consult a book on linear algebra for the proof of this solution.

that in order to solve the above system we had to find the inverse of a 2x2 matrix. In this case, it was quick to find this inverse because the coefficient matrix was small. But for larger matrices, finding the inverse using MATLAB may take more time. Therefore, it is advised to use another method to solve the above system. One such method that is rather fast in execution is called Gaussian elimination. This method is already implemented in MATLAB as matrix division using the backslash operator "\". Here is the solution of the above system again using Gaussian elimination and the backslash operator:

```
>> A = [1 -3 ; 4 6]

A =

      1      -3
      4       6

>> b = [5 ; 3]

b =

      5
      3

>> x = A\b

x =

    2.1667
   -0.9444
```

It is clear that we obtain exactly the same solution as before but with more speed on the part of MATLAB. The use of Gaussian elimination along with the backslash operator in MATLAB is greatly recommended for solving large systems of algebraic linear simultaneous equations. As an example, let us solve the following system of five algebraic linear simultaneous equations:

$$2x_1 - 4x_2 - x_3 + 3x_4 - x_5 = 3$$
$$x_1 + x_2 - 2x_3 + x_5 = 6$$
$$- x_1 - 3x_2 + x_4 + 3x_5 = -4$$
$$3x_1 - x_2 - x_3 + 4x_4 - x_5 = 1$$
$$x_1 + x_2 - x_3 + 2x_4 = 5$$

First, we re-write the above system as a matrix equation as follows:

$$\begin{bmatrix} 2 & -4 & -1 & 3 & -1 \\ 1 & 1 & -2 & 0 & 1 \\ -1 & -3 & 0 & 1 & 3 \\ 3 & -1 & -1 & 4 & -1 \\ 1 & 1 & -1 & 2 & 0 \end{bmatrix} \begin{Bmatrix} x_1 \\ x_2 \\ x_3 \\ x_4 \\ x_5 \end{Bmatrix} = \begin{Bmatrix} 3 \\ 6 \\ -4 \\ 1 \\ 5 \end{Bmatrix}$$

The following are the MATLAB commands using Gaussian elimination along with the solution:

```
>> A = [2 -4 -1 3 -1 ; 1 1 -2 0 1 ; -1 -3 0
1 3 ; 3 -1 -1 4 -1 ; 1 1 -1 2 0]

A =

    2    -4    -1     3    -1
    1     1    -2     0     1
   -1    -3     0     1     3
    3    -1    -1     4    -1
    1     1    -1     2     0

>> b = [3 ; 6 ; -4 ; 1 ; 5]

b =
```

```
      3
      6
     -4
      1
      5

>> x = A\b

x =

   -4.3571
    0.3571
   -6.4286
    1.2857
   -2.8571
```

It is seen from the above result that five real solutions are obtained for the above system of linear equations.

Systems of simultaneous nonlinear algebraic equations can also be solved in MATLAB but these are significantly more difficult to solve. The reason is that there are no direct commands to solve these complicated systems in MATLAB. One will have to use the MATLAB Optimization Toolbox to find certain functions for the solution of these types of equations – for example the command `fsolve` in this toolbox will solve a system of nonlinear equations. Another option is to use the MATLAB Symbolic Math Toolbox to solve such nonlinear systems - using the MATLAB command `solve`. This option will be illustrated in detail below.

Solving Equations with the MATLAB Symbolic Math Toolbox

The MATLAB Symbolic Math Toolbox can be used to solve algebraic equations in MATLAB. In addition, systems of linear and nonlinear algebraic equations can also be solved using this toolbox. There are special commands in this toolbox for solving equations – in particular the MATLAB command `solve`. Please note that the command `roots` that was discussed earlier in this chapter cannot be used for the symbolic solution of algebraic equations. It needs to be replaced with the `solve` command. The use of the MATLAB command `solve` will be illustrated in this section with several examples.

Consider the following linear algebraic symbolic equation that we need to solve for the variable x in terms of the constants a and b:

$$ax + b = 0$$

We will use the MATLAB command `solve` to solve the above equation. Note that we cannot use the trick of writing it as a polynomial and use the `roots`[38] command. Here is the needed format to use the `solve` command in order to solve the above equation:

```
>> syms x
>> syms a
>> syms b
>> x = solve('a*x+b = 0')

x =

-b/a
```

It is noted that the correct solution is obtained which is $x = -\dfrac{b}{a}$. Next, let us solve the following quadratic equation for x in terms of the constants a, b, and c:

---

[38] The `roots` command can only be used with numerical computations.

$$ax^2 + bx + c = 0$$

The following is the correct format of the `solve` command in order to obtain the two solutions of the above nonlinear equation:

```
>> syms x
>> syms a
>> syms b
>> syms c
>> x = solve('a*x^2 + b*x + c = 0')

x =

 1/2/a*(-b+(b^2-4*a*c)^(1/2))
 1/2/a*(-b-(b^2-4*a*c)^(1/2))
```

It is clear that we obtain the correct two solutions[39] of the quadratic formula. Let us now solve another nonlinear equation not involving a polynomial. Consider the following equation:

$$e^x = \sin x$$

The above equation is a highly nonlinear equation in x. Below is the needed format of the `solve` command to find this solution:

```
>> syms x
>> x = solve('exp(x)-sin(x) = 0')

x =

.36270205612105111082838863639609-
1.1337459194137525371167081219057*i
```

---

[39] We can use the MATLAB command `pretty` to display the solution in a nice way but this will not be done here. The correct use of the `pretty` command in the example above would be `pretty(x)`.

It is clear from the above that we obtained one complex solution of the above equation. The resulting solution is approximated as $0.363 - 1.13i$. At this step, one may use the MATLAB command `format short` to display the solution using four decimal digits only.

Next, we will consider systems of equations. Consider the following system of linear simultaneous algebraic equations:

$$ax - by = c$$
$$2ax + b = 9c$$

The above system of linear equations can also be solved for the variables x and y in terms of the constants a, b, and c using the MATLAB `solve` command as follows:

```
>> syms x
>> syms y
>> syms a
>> syms b
>> syms c
>> [x,y] = solve('a*x - b*y - c = 0', '2*a*x
+ b - 9*c = 0')

x =

-1/2/a*(b-9*c)



y =

-1/2*(-7*c+b)/b
```

It is clear that two solutions are obtained for x and y in terms of a, b, and c as shown above. Our final example will illustrate the solution of a nonlinear system of simultaneous algebraic equations. Consider the following system:

$$x^3 - xy = -3$$
$$x^2 - 2y^2 = 5$$

The above system is a highly nonlinear system. Its solution using the MATLAB command solve is illustrated below:

```
>> syms x
>> syms y
>> [x,y] = solve('x^3 - x*y + 3 = 0', 'x^2 -
2*y^2 - 5 = 0')

x =

1.1146270688593392975541180081115+1.36934983
2630038204073932908507 4*i

.369793706249890017733017779583 73+1.06182311
9513668701159057314663 1*i
 -
1.4844207751092293152871357876952+.282946327
3029445103261353459655 8*i
 -1.4844207751092293152871357876952-
.2829463273029445103261353459655 8*i
  .369793706249890017733017779583 73-
1.0618231195136687011590573146631*i
   1.1146270688593392975541180081115-
1.3693498326300382040739329085074*i


y =
```

```
.43988651431894061086040911013399+1.73489563
84245835120290450101938*i
 -.11319578028438869117106079077457-
1.73440876400289387166691238072308*i
   .17330926596544808031065168064 07-
1.211739615160225891007650174 8134*i

.17330926596544808031065168064 07+1.211739615
160225891007650174 8134*i
 -
.11319578028438869117106079077457+1.73440876
400289387166691238072308*i
  .43988651431894061086040911013399-
1.73489563842458351202904 50101938*i
```

It is clear from the above that we obtain six complex solutions for the above nonlinear system. Finally, one may use the MATLAB command format short to display the six solutions above using four decimal digits only.

In the next chapter, we will present an introduction to calculus using MATLAB.

Exercises

Solve all the exercises using MATLAB. All the needed MATLAB commands for these exercises were presented in this chapter. Note that Exercises 7-11 require the use of the MATLAB Symbolic Math Toolbox.

1. Solve the following linear algebraic equation for the variable x. Use the roots command.

    $$3x + 5 = 0$$

2. Solve the following quadratic algebraic equation for the variable x. Use the `roots` command.

$$x^2 + x + 1 = 0$$

3. Solve the following algebraic equation for x. Use the `roots` command.

$$3x^4 - 2x^3 + x - 3 = 0$$

4. Solve the following system of linear simultaneous algebraic equations for the variables x and y. Use the inverse matrix method.

$$3x + 5y = 9$$
$$4x - 7y = 13$$

5. In Exercise 4 above, solve the same linear system again but using the Gaussian elimination method with the backslash operator.

6. Solve the following system of linear simultaneous algebraic equations for the variables x, y, and z. Use Gaussian elimination with the backslash operator.

$$2x - y + 3z = 5$$
$$4x + 5z = 12$$
$$x + y + 2z = -3$$

7. Solve the following linear algebraic equation for the variable x in terms of the constant a.

$$2x + a = 5$$

8. Solve the following quadratic algebraic equation for the variable x in terms of the constants a and b.

$$x^2 + ax + b = 0$$

9. Solve the following nonlinear equation for the variable x.

$$2e^x + 3\cos x = 0$$

10. Solve the following system of linear simultaneous algebraic equations for the variables x and y in terms of the constant c.

$$2x - 3cy = 5$$
$$cx + 2y = 7$$

11. Solve the following system of nonlinear simultaneous algebraic equations for the variables x and y.

$$3x^2 - 2x + y = 7$$
$$xy + x = 5$$

11. Beginning Calculus

In this last chapter, we present an introduction to calculus using MATLAB. In order to perform such operations in MATLAB as differentiation and integration[40] of mathematical functions, it is recommended that we use the MATLAB Symbolic Math Toolbox. Therefore, this whole chapter will utilize this toolbox in introducing calculus to MATLAB users.

Beginning Calculus with the MATLAB Symbolic Math Toolbox

First, we will study the differentiation and integration of mathematical functions in MATLAB. Let us first introduce the MATLAB command `inline`[41] with which we can define a mathematical function in MATLAB. For example let us define the following function in MATLAB.

$$f(x) = 2x^3 - 5$$

Here are the needed MATLAB commands to define the above function:

```
>> f = inline('2*x^3 - 5', 'x')

f =

     Inline function:
     f(x) = 2*x^3 - 5

>> f(2)
```

---

[40] See a book on calculus for details about these mathematical operations.
[41] The `inline` command is part of the main MATLAB package, not the Symbolic Math Toolbox.

```
ans =

    11
```

The above example defines the function in MATLAB using the `inline` command. It was followed above by a numerical evaluation of the function at $x = 2$. Note that the definition of the function must be made only once while its evaluation can be executed several times. Here is another evaluation of the above function at $x = -1$.

```
>> f(-1)

ans =

    -7
```

Let us now differentiate (i.e. find the derivative) of the above function with respect to x using the MATLAB command `diff`. This is performed below as follows:

```
>> syms x
>> diff(f(x),x)

ans =

6*x^2
```

Note that we obtained the correct derivative of the function which is $6x^2$. Note also that we had to use the `syms` command to declare that x was a symbolic variable. Let us now define and differentiate the following mathematical function in MATLAB:

$$g(y) = 2 - 4\sin(\pi y)$$

The above function is defined and differentiated with respect to y as follows in MATLAB:

```
>> syms y
>> g = inline ('2 - 4*sin(pi*y)', 'y')

g =

    Inline function:
    g(y) = 2 - 4*sin(pi*y)

>> diff(g(y),y)

ans =

-4*cos(pi*y)*pi
```

In the above example, we obtain the correct derivative which is $-4\pi\cos(\pi y)$. Let us find the integral of the above function using the MATLAB command int as follows:

```
>> int(g(y))

ans =

2*y+4/pi*cos(pi*y)
```

The correct integral which is $2y + \dfrac{4}{\pi}\cos(\pi y)$ is obtained in the example above. The above integral was an indefinite integral. However, we can evaluate a definite integral like $\int_{1}^{2} g(y)dy$ using the same int command as follows (modified with some additional arguments):

```
> int(g(y),1,2)

ans =

2*(pi+4)/pi
```

The correct answer was obtained above which is $\dfrac{2(\pi+4)}{\pi}$.
We can also evaluate this symbolic answer numerically using the `double` command as follows:

```
>> double(ans)

ans =

    4.5465
```

We can also use MATLAB to evaluate second and higher derivatives by repeated uses of the `diff` command (or by using certain arguments). Also, we can use MATLAB to evaluate double and triple integrals by repeated uses of the `int` command (or by using certain arguments). However, these exercises are straightforward and will be left to the reader.

We can also use MATLAB to evaluate limits of mathematical functions using the MATLAB command `limit`. For example, let us evaluate the following limit:

$$\lim_{x\to 0}\frac{\sin x}{x}$$

Here is the needed MATLAB command to evaluate the above limit in MATLAB:

```
>> limit(sin(x)/x,x,0)
```

```
ans =
```

```
1
```

Note in the above example that we entered the mathematical function directly in the arguments of the `limit` command without the use of the `inline` command. The same thing can also be done with the `diff` and `int` commands that were used previously for differentiation and integration, respectively.

Let us evaluate the following limit in MATLAB:

$$\lim_{x \to \infty} \frac{2x^3 - 3x^2 + x + 5}{4x^3 - 7}$$

The following is the needed MATLAB command to evaluate the above limit:

```
>> limit((2*x^3-3*x^2+x+5)/(4*x^3-7),x,Inf)
```

```
ans =
```

½

Note in the above example that the correct limit value which is $\frac{1}{2}$ was obtained. Note also that we used the MATLAB symbol `Inf` to represent infinity.

We can also use MATLAB to obtain the Taylor series expansion of a mathematical function using the `taylor` command. For example, here are the commands needed to obtain the Taylor series expansions of the two functions $\sin x$ and $\ln(x+1)$:

```
>> syms x
>> taylor(sin(x),x,7)
```

```
ans =

x-1/6*x^3+1/120*x^5

>> taylor(log(x+1),x,7)

ans =

x-1/2*x^2+1/3*x^3-1/4*x^4+1/5*x^5-1/6*x^6
```

In the above two expansions, both series were evaluated up to seven terms. If the above expressions are to be evaluated at a certain point and their sum or product is needed, one can use the MATLAB commands sum[42] and prod for this purpose. One can also use the MATLAB command symsum to sum a sequence of numbers symbolically. Here is a series that needs to be summed followed by the needed MATLAB commands:

$$\sum_{1}^{n}\left(\frac{1}{k+1}-\frac{1}{k}\right)$$

```
>> syms n
>> syms k
>> symsum(1/(k+1) - 1/k,1,n)

ans =

1/(n+1)-1
```

---

[42] See the chapter on vectors for more details about this command.

The correct answer for the sum is obtained which is

$$\frac{1}{n+1} - 1.$$

There are other numerous commands available in the MATLAB Symbolic Math Toolbox that can be used to manipulate mathematical expressions in symbolic form. For example, some of these commands are expand, simplify, simple, factor, collect, gradient, and subs[43]. However, the use of these commands will not be illustrated in this introductory book.

Finally, we will illustrate the use of MATLAB in solving a simple ordinary differential equation[44] using the command dsolve. Let us solve the following initial value ordinary differential equation in x:

$$\frac{dy}{dx} = x + y \qquad , \qquad y(0) = 1$$

Below is the needed MATLAB command to solve the above equation for y as a function of x:

```
>> dsolve('Dy = x + y', 'y(0)=1')

ans =

-x+exp(t)*(x+1)
```

In the above example, the correct function is obtained using MATLAB. Higher order ordinary differential equations can also be used by repeated use of the dsolve command. Our final note is

---

[43] One of these commands has been used in a previous chapter.
[44] A differential equation is an equation that has one or more derivatives as the unknown(s). Actually, the unknowns are functions that need to be determined.

that there is a special command in MATLAB called ode45[45] that can be used for the numerical solution of ordinary differential equations but its use is beyond the scope of this book.

Exercises

Solve all the exercises using MATLAB. All the needed MATLAB commands for these exercises were presented in this chapter. Note that most of these exercises (1-15) require the use of the MATLAB Symbolic Math Toolbox.

1. Define the following mathematical function in MATLAB using the inline command:

$$f(x) = 3x^2 + x - 1$$

2. In Exercise 1 above, evaluate the function f at $x = 1$.

3. In Exercise 1 above, evaluate the function f at $x = -2$.

4. In Exercise 1 above, differentiate the function f with respect to x.

5. Define the following mathematical function in MATLAB using the inline command:

$$g(y) = 2\sin(\pi y) + 3y\cos(\pi y)$$

6. In Exercise 5 above, differentiate the function g with respect to y.

---

[45] There is a number of such commands in MATLAB called ODE Solvers for the numerical solution of ordinary differential equations. These commands are part of the main MATLAB package and are not associated with the MATLAB Symbolic Math Toolbox

7. In Exercise 5 above, find the indefinite integral of the function g.

8. In Exercise 5 above, find the value of the following definite integral

$$\int_0^1 g(y)dy$$

9. In Exercise 8 above, evaluate the value obtained numerically using the double command.

10. Evaluate the following limit in MATLAB:

$$\lim_{x \to 0}(\sin x + \cos x)$$

11. Evaluate the following limit in MATLAB:

$$\lim_{x \to \infty} \frac{x^2 + x + 1}{3x^2 - 2}$$

12. Find the Taylor series expansion for the function $\cos x$ up to eight terms.

13. Find the Taylor series expansion for the function e^x up to nine terms.

14. Evaluate the following sum symbolically using the symsum command:

$$\sum_1^n \frac{1}{k}$$

15. Solve the following initial value ordinary differential equation using the dsolve command:

$$\frac{dy}{dx} = xy - \sin x + 3 \qquad , \qquad y(0) = 0$$

Solutions to Exercises

1. Introduction

1. Perform the operation 3*4+6. The order of the operations will be discussed in subsequent chapters.

    ```
    >> 3*4+6

    ans =

        18
    ```

2. Perform the operation $\cos(5)$. The value of 5 is in radians.

    ```
    >> cos(5)

    ans =

        0.2837
    ```

3. Perform the operation $3\sqrt{6+x}$ for $x = 4$.

    ```
    >> x = 4

    x =

            4

    >> 3*sqrt(6+x)

    ans =

        9.4868
    ```

4. Assign the value of 5.2 to the variable y.

```
>> y = 5.2

y =

    5.2000
```

5. Assign the values of 3 and 4 to the variables x and y, respectively, then calculate the value of z where $z = 2x - 7y$.

```
>> x = 3;
>> y = 4;
>> z = 2*x - 7*y

z =

    -22
```

6. Obtain help on the inv command.

```
>> help inv
 INV    Matrix inverse.
    INV(X)  is  the  inverse  of  the  square
matrix X.
    A  warning  message  is  printed  if  X  is
badly scaled or
    nearly singular.

    See  also  slash,  pinv,  cond,  condest,
lsqnonneg, lscov.

    Overloaded  functions  or  methods  (ones
with the same name in other directories)
        help sym/inv.m

    Reference page in Help browser
        doc inv
```

7. Generate the following matrix: A

$$A = \begin{bmatrix} 1 & 0 & 2 & -3 \\ 0 & 5 & 2 & 2 \\ 1 & 2 & 3 & 4 \\ -2 & 0 & 1 & 3 \end{bmatrix}$$

```
>> A = [1 0 2 -3 ; 0 5 2 2 ; 1 2 3 4 ; -2 0
1 3]

A =

    1        0        2       -3
    0        5        2        2
    1        2        3        4
   -2        0        1        3
```

8. Generate the following vector b

$$b = \begin{Bmatrix} 1 \\ 2 \\ 3 \\ 4 \end{Bmatrix}$$

```
>> b = [1 ; 2 ; 3 ; 4]

b =

    1
    2
    3
    4
```

9. Evaluate the vector c where $\{c\} = [A]\{b\}$ where A is the matrix given in Exercise 7 above and b is the vector given in Exercise 8 above.

```
>> c = A*b
```

```
c =

    -5
    24
    30
    13
```

10. Solve the following system of simultaneous algebraic equation using Gaussian elimination.

$$\begin{bmatrix} 5 & 2 \\ 1 & 3 \end{bmatrix} \begin{Bmatrix} x_1 \\ x_2 \end{Bmatrix} = \begin{Bmatrix} 3 \\ -1 \end{Bmatrix}$$

```
>> A = [ 5 2 ; 1 3]

A =

    5        2
    1        3

>> b = [3 ; -1]

b =

    3
   -1

>> x = A\b

x =

    0.8462
   -0.6154
```

11. Solve the system of simultaneous algebraic equations of Exercise 10 above using matrix inversion.

```
>> x = inv(A)*b
```

```
x =

    0.8462
   -0.6154
```

12. Generate the following matrix X:

$$X = \begin{bmatrix} 1 & 0 & 6 \\ 1 & 2 & 3 \\ 4 & 5 & -2 \end{bmatrix}$$

```
>> X = [1 0 6 ; 1 2 3 ; 4 5 -2]

X =

    1        0        6
    1        2        3
    4        5       -2
```

13. Extract the sub-matrix in rows 2 to 3 and columns 1 to 2 of the matrix X in Exercise 12 above.

```
>> X(2:3,1:2)

ans =

    1        2
    4        5
```

14. Extract the second column of the matrix X in Exercise 12 above.

```
>> X(1:3,2)

ans =

    0
    2
    5
```

15. Extract the first row of the matrix X in Exercise 12 above.

```
>> X(1,1:3)

ans =

     1    0    6
```

16. Extract the element in row 1 and column 3 of the matrix X in Exercise 12 above.

```
>> X(1,3)

ans =

     6
```

17. Generate the row vector x with integer values ranging from 1 to 9.

```
>> x = [1 2 3 4 5 6 7 8 9]

x =

     1    2    3    4    5    6    7
     8    9
```

18. Plot the graph of the function $y = x^3 - 2$ for the range of the values of x in Exercise 17 above.

```
>> y = x.^3 - 2

y =

    -1    6    25    62    123    214    341
    510    727

>> plot(x,y)
```

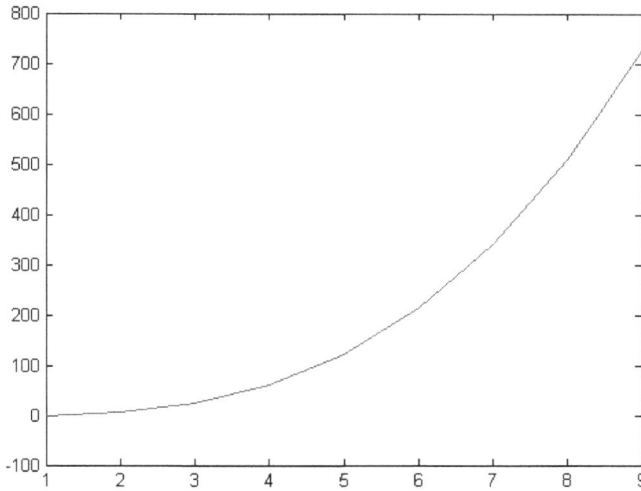

19. Generate a 4 x 4 magic square. What is the total of each row, column, and diagonal in this matrix.

```
>> magic(4)

ans =

    16     2     3    13
     5    11    10     8
     9     7     6    12
     4    14    15     1
```

The sum of each row, column, or diagonal is 34.

2. Arithmetic Operations

1. Perform the addition operation 7+9.

    ```
    >> 7+9
    ```

```
ans =

    16
```

2. Perform the subtraction operation 16-10.

```
>> 16-10

ans =

     6
```

3. Perform the multiplication operation 2*9.

```
>> 2*9

ans =

    18
```

4. Perform the division operation 12/3.

```
>> 12/3

ans =

     4
```

5. Perform the division operation 12/5.

```
>> 12/5

ans =

    2.4000
```

6. Perform the exponentiation operation 3^5.

```
>> 3^5
```

```
ans =

    243
```

7. Perform the exponentiation operation 3*(-5).

```
>> 3^(-5)

ans =

    0.0041
```

8. Perform the exponentiation operation (-3)^5.

```
>> (-3)^5

ans =

   -243
```

9. Perform the exponentiation operation -3^5.

```
>> -3^5

ans =

   -243
```

10. Compute the value of $\dfrac{2\pi}{3}$.

```
>> 2*pi/3

ans =

    2.0944
```

11. Obtain the value of the smallest number that can be handled by MATLAB.

```
>> eps

ans =

   2.2204e-016
```

12. Perform the multiple operations 5+7-15.

```
>> 5+7-15

ans =

    -3
```

13. Perform the multiple operations (6*7)+4.

```
>> 6*7 +4

ans =

    46
```

14. Perform the multiple operations 6*(7+4).

```
>> 6*(7+4)

ans =

    66
```

15. Perform the multiple operations 4.5 + (15/2).

```
>> 4.5 + 15/2

ans =

    12
```

16. Perform the multiple operations (4.5 + 15)/2.

```
>> (4.5 + 15)/2

ans =

    9.7500
```

17. Perform the multiple operations $(15 - 4 + 12)/5 - 2*(7^4)/100$.

```
>> (15-4+12)/5 - 2*(7^4)/100

ans =

  -43.4200
```

18. Perform the multiple operations $(15 - 4) + 12/5 - (2*7)^4/100$.

```
>> (15-4) + 12/5 - (2*7)^4/100

ans =

  -370.7600
```

19. Define the number 2/3 as a symbolic number.

```
>> sym(2/3)

ans =

2/3
```

20. Perform the fraction addition $(2/3) + (3/4)$ numerically.

```
>> 2/3 + 3/4

ans =

    1.4167
```

21. Perform the fraction addition $(2/3) + (3/4)$ symbolically.

```
>> sym((2/3)+(3/4))

ans =

17/12
```

3. Variables

1. Perform the operation $2*3+7$ and store the result in the variable w.

```
>> 2*3+7

ans =

    13
```

2. Define the three variables a, b, and c equal to 4, -10, and 3.2, respectively.

```
>> a = 4

a =

        4

>> b = -10

b =

    -10

>> c = 3.2

c =

    3.2000
```

3. Define the two variables y and Y equal to 10 and 100. Are the two variables identical?

```
>> y = 10

y =

    10

>> Y = 100

Y =

    100
```

The two variables are not identical.

4. Let x = 5.5 and y = -2.6. Calculate the value of the variable z = 2x-3y.

```
x =

    5.5000

>> y = -2.6

y =

    -2.6000

>> z = 2*x - 3*y

z =

    18.8000
```

5. In Exercise 4 above, calculate the value of the variable w = 3y - z + x/y.

```
>> w = 3*y - z + x/y
```

```
w =

    -28.7154
```

6. Let r = 6.3 and s = 5.8. Calculate the value of the variable final defined by final = r + s - r*s.

```
>> r = 6.3

r =

     6.3000

>> s = 5.8

s =

     5.8000

>> final = r + s - r*s

final =

    -24.4400
```

7. In Exercise 6 above, calculate the value of the variable this_is_the_result defined by this_is_the_result = r^2 - s^2.

```
>> this_is_the_result = r^2 - s^2

this_is_the_result =

     6.0500
```

8. Define the three variable width, Width, and WIDTH equal to 1.5, 2.0, and 4.5, respectively. Are these three variables identical?

```
>> width = 1.5
```

```
width =

    1.5000

>> Width = 2.0

Width =

    2

>> WIDTH = 4.5

WIDTH =

    4.5000
```

The three variables are not identical.

9. Write the following comment in MATLAB: `This line will not be executed.`

    ```
    % This line will not be executed.
    ```

10. Assign the value of `3.5` to the variable `s` then add a comment about this assignment on the same line.

    ```
    >> s = 3.5    % the variable s is assigned
    the value 3.5

    s =

        3.5000
    ```

11. Define the values of the variables y1 and y2 equal to 7 and 9 then perform the calculation y3 = y1 - y2/3. (Note: 2 in the formula is a subscript and should not be divided by 3).

    ```
    >> y1 = 7
    ```

```
y1 =

    7

>> y2 = 9

y2 =

    9

>> y3 = y1 - y2/3

y3 =

    4
```

12. Perform the operation $2*m - 5$. Do you get an error? Why?

```
>> 2*m - 5
??? Undefined function or variable 'm'.
```

We get an error message because the variable m is not defined.

13. Define the variables cost and profit equal to 175 and 25, respectively, then calculate the variable sale_price defined by sale_price = cost + profit.

```
>> cost = 175

cost =

    175

>> profit = 25

profit =

    25

>> sale_price = cost + profit
```

```
sale_price =

    200
```

14. Define the variable centigrade equal to 28 then calculate the variable fahrenheit defined by fahrenheit = (centigrade*9/5) + 32.

```
>> centigrade = 28

centigrade =

     28

>> fahrenheit = (centigrade*9/5) + 32

fahrenheit =

    82.4000
```

15. Use the format short and format long commands to write the values of 14/9 to four decimals and sixteen digits, respectively.

```
>> format long
>> 14/9

ans =

    1.55555555555556

>> format short
>> 14/9

ans =

    1.5556
```

16. Perform the who command to get a list of the variables stored in this session.

```
>> who

Your variables are:

WIDTH               fahrenheit              x
Width                final                  y
Y                    profit                 y1
a                    r                      y2
ans                  s                      y3
b                    sale_price             z
c              this_is_the_result     centigrade
w                    cost                 width
```

17. Perform the whos command to get a list of the variables stored in this session along with their details.

```
>> whos
  Name         Size      Bytes       Class

  WIDTH        1x1        8        double array
  Width        1x1        8        double array
  Y            1x1        8        double array
  a            1x1        8        double array
  ans          1x1        8        double array
  b            1x1        8        double array
  c            1x1        8        double array
  centigrade   1x1        8        double array
  cost         1x1        8        double array
  fahrenheit   1x1        8        double array
  final        1x1        8        double array
  profit       1x1        8        double array
  r            1x1        8        double array
  s            1x1        8        double array
  sale_price   1x1        8        double array
  this_is_the_result 1x1  8        double array
  w            1x1        8        double array
  width        1x1        8        double array
```

```
x                1x1              8        double array
y                1x1              8        double array
y1               1x1              8        double array
y2               1x1              8        double array
y3               1x1              8        double array
z                1x1              8        double array
```

Grand total is 24 elements using 192 bytes

18. Clear all the variables stored in this session by using the `clear` command.

```
>> clear
>>
```

19. Calculate both the area and perimeter of a rectangle of sides 5 and 7. No units are used in this exercise.

```
a =

     5

>> b = 7

b =

     7

>> area = a*b

area =

    35

>> perimeter = 2*(a+b)

perimeter =

    24
```

20. Calculate both the area and perimeter of a circle of radius 6.45. No units are used in this exercise.

```
>> r = 6.45

r =

    6.4500

>> area = pi*r^2

area =

  130.6981

>> perimeter = 2*pi*r

perimeter =

   40.5265
```

21. Define the symbolic variables x and z with values 4/5 and 14/17.

```
>> x = 4/5

x =

    0.8000

>> sym(x)

ans =

4/5


>> z = 14/17

z =
```

```
      0.8235

>> sym(z)

ans =

14/17
```

22. In Exercise 21 above, calculate symbolically the value of the symbolic variable y defined by $y = 2x - z$.

```
>> y = sym(2*x - z)

y =

66/85
```

23. Calculate symbolically the area of a circle of radius 2/3 without obtaining a numerical value. No units are used in this exercise.

```
>> radius1 = 2/3

radius1 =

      0.6667

>> radius1 = sym(radius1)

radius1 =

2/3


>> area = pi*radius1^2

area =

4/9*pi
```

24. Calculate symbolically the volume of a sphere of radius 2/3 without obtaining a numerical value. No units are used in this exercise.

```
>> radius2 = 2/3

radius2 =

    0.6667

>> radius2 = sym(radius2)

radius2 =

2/3


>> volume = (4/3)*pi*radius2^3

volume =

32/81*pi
```

25. In Exercise 23 above, use the `double` command to obtain the numerical value of the answer.

```
>> double(area)

ans =

    1.3963
```

26. In Exercise 24 above, use the `double` command to obtain the numerical value of the answer.

```
>> double(volume)

ans =

    1.2411
```

27. Define the symbolic variables y and `date` without assigning any numerical values to them.

```
>> y = sym('y')

y =

y


>> date = sym('date')

date =

date
```

4. Mathematical Functions

1. Compute the square root of 10.

```
>> sqrt(10)

ans =

    3.1623
```

2. Compute the factorial of 7.

```
>> factorial(7)

ans =

        5040
```

3. Compute the cosine of the angle 45 where 45 is in radians.

```
>> cos(45)
```

```
ans =

    0.5253
```

4. Compute the cosine of the angle 45 where 45 is in degrees.

```
>> cos(45*pi/180)

ans =

    0.7071
```

5. Compute the sine of the angle of 45 where 45 is in degrees.

```
>> sin(45*pi/180)

ans =

    0.7071
```

6. Compute the tangent of the angle 45 where 45 is in degrees.

```
>> tan(45*pi/180)

ans =

    1.0000
```

7. Compute the inverse tangent of 1.5.

```
>> atan(1.5)

ans =

    0.9828
```

The above result is in radians.

8. Compute the tangent of the angle $\dfrac{3\pi}{2}$. Do you get an error? Why?

```
>> tan(3*pi/2)

ans =

   5.4437e+015
```

We get a very large number approaching infinity because the tanget function is not define at $\dfrac{3\pi}{2}$.

9. Compute the value of exponential function e^3.

```
>> exp(3)

ans =

   20.0855
```

10. Compute the value of the natural logarithm $\ln 3.5$.

```
>> log(3.5)

ans =

   1.2528
```

11. Compute the value of the logarithm $\log_{10} 3.5$.

```
>> log10(3.5)

ans =

   0.5441
```

12. Use the MATLAB rounding function round to round the value of 2.43.

```
>> round(2.43)
```

```
ans =

    2
```

13. Use the MATLAB remainder function rem to obtain the remainder when dividing 5 by 4.

```
>> rem(5,4)

ans =

    1
```

14. Compute the absolute value of -3.6.

```
>> abs(-3.6)

ans =

    3.6000
```

15. Compute the value of the expression $1.5 - 2\sqrt{6.7/5}$.

```
>> 1.5 - 2*sqrt(6.7/5)

ans =

    -0.8152
```

16. Compute the value of $\sin^2 \pi + \cos^2 \pi$.

```
>> sin(pi)^2 + cos(pi)^2

ans =

    1
```

17. Compute the value of $\log_{10} 0$. Do you get an error? Why?

```
>> log10(0)
Warning: Log of zero.
> In log10 at 17

ans =

  -Inf
```

We get minus infinity because the logarithmic function is not defined at zero.

18. Let $x = \dfrac{3\pi}{2}$ and $y = 2\pi$. Compute the value of the expression $2\sin x \cos y$.

```
>> x = 3*pi/2

x =

      4.7124

>> y = 2*pi

y =

      6.2832

>> 2*sin(x)*cos(y)

ans =

     -2
```

19. Compute the value of $\sqrt{45}$ symbolically and simplify the result.

```
>> sym(sqrt(45))

ans =
```

```
sqrt(45)
```

```
>> simplify(ans)
```

```
ans =
```

```
3*5^(1/2)
```

20. Compute the value of $\sqrt{45}$ numerically.

```
>> double(ans)
```

```
ans =
```

```
    6.7082
```

21. Compute the sine of the angle 45 (degrees) symbolically.

```
>> sym(sin(45*pi/180))
```

```
ans =
```

```
sqrt(1/2)
```

22. Compute the cosine of the angle 45 (degrees) symbolically.

```
> sym(cos(45*pi/180))
```

```
ans =
```

```
sqrt(1/2)
```

23. Compute the tangent of the angle 45 (degrees) symbolically.

```
>> sym(tan(45*pi/180))
```

```
ans =
```

1

24. Compute the value of $e^{\pi/2}$ symbolically.

```
>> sym(exp(pi/2))

ans =

5416116035097439*2^(-50)
```

25. Compute the value of $e^{\pi/2}$ numerically.

```
>> double(ans)

ans =

    4.8105
```

5. Complex Numbers

1. Compute the square root of -5.

```
>> sqrt(-5)

ans =

        0 + 2.2361i
```

2. Define the complex number $4 - 3\sqrt{-8}$.

```
>> 4-3*sqrt(-8)

ans =

    4.0000 - 8.4853i
```

3. Define the two complex numbers with variables x and y where $x = 2 - 6i$ and $y = 4 + 11i$.

```
>> x = 2-6i

x =

   2.0000 -  6.0000i

>> y=4+11i

y =

   4.0000 +11.0000i
```

4. In Exercise 3 above, perform the addition and subtraction operations $x + y$ and $x - y$.

```
>> x+y

ans =

   6.0000 + 5.0000i

>> x-y

ans =

  -2.0000 -17.0000i
```

5. In Exercise 3 above, perform the multiplication and division operations $x\,y$ and $\dfrac{x}{y}$.

```
>> x*y

ans =

  74.0000 - 2.0000i

>> x/y
```

```
ans =

   -0.4234 - 0.3358i
```

6. In Exercise 3 above, perform the exponentiation operations x^4 and y^{-3}.

```
>> x^4

ans =

   4.4800e+002 +1.5360e+003i

>> y^(-3)

ans =

   -5.3979e-004 +3.1229e-004i
```

7. In Exercise 3 above, perform the multiple operations $4x - 3y + 9$.

```
>> 4*x-3*y+9

ans =

   5.0000 -57.0000i
```

8. In Exercise 3 above, perform the multiple operations $ix - 2y - 1$.

```
>> i*x-2*y-1

ans =

   -3.0000 -20.0000i
```

9. Compute the magnitude of the complex number $3 - 5i$.

```
>> abs(3-5i)

ans =

    5.8310
```

10. Compute the angle of the complex number $3-5i$ in radians.

```
>> angle(3-5i)

ans =

   -1.0304
```

11. Compute the angle of the complex number $3-5i$ in degrees.

```
>> angle(3-5i)*180/pi

ans =

   -59.0362
```

12. Extract the real and imaginary parts of the complex number $3-5i$.

```
>> real(3-5i)

ans =

      3

>> imag(3-5i)

ans =

   -5
```

13. Obtain the complex conjugate of the complex number $3-5i$.

```
>> conj(3-5i)
```

```
ans =

    3.0000 + 5.0000i
```

14. Compute the sine, cosine, and tangent functions of the complex number $3-5i$.

```
>> sin(3-5i)

ans =

   10.4725 +73.4606i

>> cos(3-5i)

ans =

  -73.4673 +10.4716i

>> tan(3-5i)

ans =

   -0.0000 - 0.9999i
```

15. Compute e^{3-5i} and $\ln(3-5i)$.

```
>> exp(3-5i)

ans =

    5.6975 +19.2605i

>> log(3-5i)

ans =

    1.7632 - 1.0304i
```

16. Compute the values of $\sin\dfrac{\pi i}{2}$, $\cos\dfrac{\pi i}{2}$, and $e^{\pi i/2}$.

```
>> sin(pi*i/2)

ans =

        0 + 2.3013i

>> cos(pi*i/2)

ans =

    2.5092

>> exp(pi*i/2)

ans =

    0.0000 + 1.0000i
```

17. Compute the value of $(3+4i)^{(2-i)}$.

```
>> (3+4i)^(2-i)

ans =

    61.3022 +15.3369i
```

18. Obtain $\sqrt{-13}$ symbolically.

```
>> sym(sqrt(-13))

ans =

(0)+(sqrt(13))*i
```

19. Obtain the magnitude of the complex number $3-5i$ symbolically.

```
>> sym(abs(3-5i))

ans =

sqrt(34)
```

20. Obtain the angle of the complex number $3-5i$ symbolically. Make sure that you use the double command at the end.

```
>> sym(angle(3-5i))

ans =

-4640404691986088*2^(-52)


>> double(ans)

ans =

    -1.0304
```

21. Obtain the cosine function of the complex number $3-5i$ symbolically. Make sure that you use the double command at the end.

```
>> sym(cos(3-5i))

ans =

(-5169801091137321*2^(-
46))+(5894962905280379*2^(-49))*i


>> double(ans)

ans =

    -73.4673 +10.4716i
```

6. Vectors

1. Store the vector [2 4 -6 0] in the variable w.

   ```
   >> w = [2 4 -6 0]

   w =

        2      4      -6      0
   ```

2. In Exercise 1 above, extract the second element of the vector w.

   ```
   > w(2)

   ans =

        4
   ```

3. In Exercise 1 above, generate the vector z where $z = \dfrac{\pi}{2} w$.

   ```
   >> z = pi*w/2

   z =

        3.1416      6.2832      -9.4248                  0
   ```

4. In Exercise 3 above, extract the fourth element of the vector z.

   ```
   >> z(4)

   ans =

        0
   ```

5. In Exercise 3 above, extract the first three elements of the vector z.

   ```
   >> z(1:3)
   ```

```
ans =

    3.1416      6.2832     -9.4248
```

6. In Exercise 3 above, find the length of the vector z.

```
>> length(z)

ans =

    4
```

7. In Exercise 3 above, find the total sum of the values of the elements of the vector z.

```
>> sum(z)

ans =

    0
```

8. In Exercise 3 above, find the minimum and maximum values of the elements of the vector z.

```
>> min(z)

ans =

   -9.4248

>> max(z)

ans =

    6.2832
```

9. Generate a vector r with real values between 1 and 10 with an increment of 2.5.

```
>> r = (1:2.5:10)

r =

    1.0000    3.5000    6.0000    8.5000
```

10. Generate a vector s with real values of ten numbers that are equally spaced between 1 and 100.

```
>> s = linspace(1,100,10)

s =

     1     12     23     34     45     56     67
    78     89    100
```

11. Form a new vector by joining the two vectors [9 3 -2 5 0] and [1 2 -4].

```
>> a = [9 3 -2 5 0]

a =

     9     3    -2     5     0

>> b = [1 2 -4]

b =

     1     2    -4

>> c = [a b]

c =

     9     3    -2     5     0     1     2
    -4
```

12. Form a new vector by joining the vector [9 3 -2 5 0] with the number 4.

```
>> d = [a 4]

d =

     9       3      -2       5       0       4
```

13. Add the two vectors [0.2 1.3 -3.5] and [0.5 -2.5 1.0].

```
>> x = [0.2 1.3 -3.5]

x =

     0.2000     1.3000     -3.5000

>> y = [0.5 -2.5 1.0]

y =

     0.5000     -2.5000     1.0000

>> x+y

ans =

     0.7000     -1.2000     -2.5000
```

14. Subtract the two vectors in Exercise 13 above.

```
>> x-y

ans =

     -0.3000     3.8000     -4.5000
```

15. Try to multiply the two vectors in Exercise 13 above. Do you get an error message? Why?

```
>> x*y
??? Error using ==> mtimes
```

```
Inner matrix dimensions must agree.
```

We get an error because the two vectors do not have the same length.

16. Multiply the two elements in Exercise 13 above element by element.

```
>> x.*y

ans =

     0.1000    -3.2500    -3.5000
```

17. Divide the two elements in Exercise 13 above element by element.

```
>> x./y

ans =

     0.4000    -0.5200    -3.5000
```

18. Find the dot product of the two vectors in Exercise 13 above.

```
>> x*y'

ans =

    -6.6500
```

19. Try to add the two vectors [1 3 5] and [3 6]. Do you get an error message? Why?

```
>> r = [ 1 3 5 ]

r =

     1        3        5

>> s = [ 3 6 ]
```

```
s =

        3       6

>> r+s
??? Error using ==> plus
Matrix dimensions must agree.
```

We get an error message because the two vectors do not have the same length.

20. Try to subtract the two vectors in Exercise 20 above. Do you get an error message? Why?

```
>> r-s
??? Error using ==> minus
Matrix dimensions must agree.
```

We get an error message because the two vectors do not have the same length.

21. Let the vector w be defined by w = [0.1 1.3 -2.4]. Perform the operation of scalar addition 5+w.

```
>> w = [0.1 1.3 -2.4]

w =

     0.1000     1.3000    -2.4000

>> 5+w

ans =

     5.1000     6.3000     2.6000
```

22. In Exercise 22 above, perform the operation of scalar subtraction -2-w.

```
>> 2-w

ans =

    1.9000      0.7000      4.4000
```

23. In Exercise 22 above, perform the operation of scalar multiplication 1.5*w.

```
>> 1.5*w

ans =

    0.1500      1.9500      -3.6000
```

24. In Exercise 22 above, perform the operation of scalar division w/10.

```
>> w/10

ans =

    0.0100      0.1300      -0.2400
```

25. In Exercise 22 above, perform the operation 3 − 2*w/5.

```
>> 3-2*w/5

ans =

    2.9600      2.4800      3.9600
```

26. Define the vector b by b = [0 pi/3 2pi/3 pi]. Evaluate the three vectors $\sin b$, $\cos b$, and $\tan b$ (element by element).

```
>> b = [0 pi/3 2*pi/3 pi]

b =

         0    1.0472      2.0944      3.1416
```

```
>> sin(b)

ans =

            0      0.8660      0.8660      0.0000

>> cos(b)

ans =

       1.0000      0.5000     -0.5000     -1.0000

>> tan(b)

ans =

            0      1.7321     -1.7321     -0.0000
```

27. In Exercise 26 above, evaluate the vector e^b (element by element).

```
>> exp(b)

ans =

       1.0000      2.8497      8.1205     23.1407
```

28. In Exercise 26 above, evaluate the vector $\sqrt{b}$ (element by element).

```
>> sqrt(b)

ans =

            0      1.0233      1.4472      1.7725
```

29. Try to evaluate the vector 3^b. Do you get an error message? Why?

```
>> 3^b
```

```
??? Error using ==> mpower
Matrix must be square.
```

Yes, because this operation needs to be performed element by element.

30. Perform the operation in Exercise 29 above element by element?

```
>> 3.^b
```

```
ans =
```

```
    1.0000      3.1597      9.9834      31.5443
```

31. Generate a vector of 1's with a length of 4 elements.

```
> ones(1,4)
```

```
ans =
```

```
    1       1       1       1
```

32. Generate a vector of 0's with a length of 6 elements.

```
>> zeros(1,6)
```

```
ans =
```

```
    0       0       0       0       0       0
```

33. Sort the elements of the vector [0.35 -1.0 0.24 1.30 -0.03] in ascending order.

```
>> k = [0.35 -1.0 0.24 1.30 -0.03]
```

```
k =
```

```
    0.3500      -1.0000      0.2400      1.3000
-0.0300
```

```
>> sort(k)

ans =

    -1.0000      -0.0300         0.2400         0.3500
  1.3000
```

34. Generate a random permutation vector with 5 elements.

```
>> randperm(5)

ans =

     2     4     3     5     1
```

35. For the vector [2 4 -3 0 1 5 7], determine the range, mean, and median.

```
>> v = [2 4 -3 0 1 5 7]

v =

     2     4    -3     0     1     5     7

>> range(v)

ans =

    10

>> mean(v)

ans =

    2.2857

>> median(v)

ans =
```

2

36. Define the symbolic vector x = [r s t u v].

```
>> syms r s t u v
>> x = [r s t u v]

x = [ r, s, t, u, v]
```

37. In Exercise 36 above, perform the addition operation of the two vectors x and [1 0 -2 3 5] to obtain the new symbolic vector y.

```
>> r = [1 0 -2 3 5]

r =

    1     0    -2     3     5

>> y = x+r

y =

[ r+1,    s, t-2, u+3, v+5]
```

38. In Exercise 37 above, extract the third element of the vector y.

```
>> y(3)

ans =

t-2
```

39. In Exercise 37 above, perform the operation 2*x/7 + 3*y.

```
>> 2*x/7 + 3*y

ans =

[     23/7*r+3,            23/7*s,       23/7*t-6,
23/7*u+9,  23/7*v+15]
```

40. In Exercise 37 above, perform the dot products x*y' and y*x'.
Are the two results the same? Why?

```
>> x*y'

ans =

r*(1+conj(r))+s*conj(s)+t*(-
2+conj(t))+u*(3+conj(u))+v*(5+conj(v))


>> y*x'

ans =

(r+1)*conj(r)+s*conj(s)+(t-
2)*conj(t)+(u+3)*conj(u)+(v+5)*conj(v)
```

The two results are different because MATLAB treats these
symbolic variables as complex variables by default. If they were real
variables, then the results would be the same.

41. In Exercise 37 above, find the square root of the symbolic vector
x+y.

```
>> sqrt(x+y)

ans =

[    (2*r+1)^(1/2),  2^(1/2)*s^(1/2),     (2*t-
2)^(1/2),      (2*u+3)^(1/2),      (2*v+5)^(1/2)]
```

7. Matrices

1. Generate the following rectangular matrix in MATLAB. What is
the size of this matrix?

$$A = \begin{bmatrix} 3 & 0 & -2 \\ 1 & 3 & 5 \end{bmatrix}$$

```
>> A = [3 0 -2 ; 1 3 5]

A =

     3      0     -2
     1      3      5
```

A is a rectangular matrix of size 2x3.

2. In Exercise 1 above, extract the element in the second row and second column of the matrix A.

```
>> A(2,2)

ans =

     3
```

3. In Exercise 1 above, generate a new matrix B of the same size by multiplying the matrix A by the number $\dfrac{3\pi}{2}$.

```
>> B = 3*pi*A/2

B =

    14.1372          0    -9.4248
     4.7124    14.1372    23.5619
```

4. In Exercise 3 above, extract the element in the first row and third column of the matrix B.

```
>> B(1,3)

ans =
```

```
        -9.4248
```

5. In Exercise 3 above, extract the sub-matrix of the elements in common between the first and second rows and the second and third columns of the matrix B. What is the size of this new sub-matrix?

```
>> B(1:2,2:3)

ans =

            0    -9.4248
      14.1372    23.5619
```

The above sub-matrix has size 2x2.

6. In Exercise 3 above, determine the size of the matrix B using the MATLAB command size?

```
>> size(B)

ans =

      2      3
```

7. In Exercise 3 above, determine the largest of the number of rows and columns of the matrix B using the MATLAB command length?

```
>> length(B)

ans =

      3
```

8. In Exercise 3 above, determine the number of elements in the matrix B using the MATLAB command numel?

```
>> numel(B)

ans =

     6
```

9. In Exercise 3 above, determine the total sum of each column of the matrix B? Determine also the minimum value and the maximum value of each column of the matrix B?

```
>> sum(B)

ans =

    18.8496    14.1372    14.1372

>> min(B)

ans =

     4.7124          0   -9.4248

>> max(B)

ans =

    14.1372    14.1372    23.5619
```

10. Combine the three vectors [1 3 0 -4] , [5 3 1 0], and [2 2 -1 1] to obtain a new matrix of size 3x4.

```
>> a = [1 3 0 -4]

a =

     1     3     0    -4

>> b = [5 3 1 0]

b =
```

```
        5       3       1       0

>> c = [2 2 -1 1]

c =

        2       2       -1      1

>> d = [a ; b ; c]

d =

        1       3       0       -4
        5       3       1       0
        2       2       -1      1
```

11. Perform the operations of matrix addition and matrix subtraction on the following two matrices:

$$R = \begin{bmatrix} 1 & 2 & 0 \\ 7 & 5 & -3 \\ 3 & 1 & 1 \end{bmatrix}, \quad S = \begin{bmatrix} 1 & 3 & -2 \\ 3 & 5 & 7 \\ 2 & 3 & 0 \end{bmatrix}$$

```
>> R = [1 2 0 ; 7 5 -3 ; 3 1 1]

R =

        1       2       0
        7       5       -3
        3       1       1

>> S = [1 3 -2 ; 3 5 7 ; 2 3 0]

S =

        1       3       -2
        3       5       7
```

```
        2       3       0

>> R+S

ans =

        2       5      -2
       10      10       4
        5       4       1

>> R-S

ans =

        0      -1       2
        4       0     -10
        1      -2       1
```

12. In Exercise 11 above, multiply the two matrices R and S element-by-element.

```
>> R.*S

ans =

        1       6       0
       21      25     -21
        6       3       0
```

13. In Exercise 11 above, divide the two matrices R and S element-by-element.

```
>> R./S
Warning: Divide by zero.

ans =

    1.0000    0.6667         0
    2.3333    1.0000   -0.4286
    1.5000    0.3333       Inf
```

Notice the division by zero (not allowed) above in the last element.

14. In Exercise 11 above, perform the operation of matrix multiplication on the two matrices R and S. Do you get an error? Why?

```
>> R*S

ans =

        7        13        12
       16        37        21
        8        17         1
```

We do not get an error because the number of columns of the first matrix is equal to the number of rows of the second matrix.

15. Add the number 5 to each element of the matrix X given below:

$$X = \begin{bmatrix} 1 & -2 & 0 & 1 \\ 2 & 3 & 6 & 2 \\ -3 & 5 & 2 & 1 \\ 5 & -2 & 4 & 4 \end{bmatrix}$$

```
>> X = [1 -2 0 1 ; 2 3 6 2 ; -3 5 2 1 ; 5 -
   2 4 4]

X =

        1       -2        0        1
        2        3        6        2
       -3        5        2        1
        5       -2        4        4

>> 5+X

ans =
```

```
    6      3      5      6
    7      8     11      7
    2     10      7      6
   10      3      9      9
```

16. In Exercise 15 above, subtract the number 3 from each element of the matrix X.

```
>> X-3

ans =

   -2     -5     -3     -2
   -1      0      3     -1
   -6      2     -1     -2
    2     -5      1      1
```

17. In Exercise 15 above, multiply each element of the matrix X by the number -3.

```
>> -3*X

ans =

   -3      6      0     -3
   -6     -9    -18     -6
    9    -15     -6     -3
  -15      6    -12    -12
```

18. In Exercise 15 above, divide each element of the matrix X by the number 2.

```
>> X/2

ans =

    0.5000    -1.0000         0    0.5000
    1.0000     1.5000    3.0000    1.0000
   -1.5000     2.5000    1.0000    0.5000
```

```
        2.5000     -1.0000     2.0000     2.0000
```

19. In Exercise 15 above, perform the following multiple scalar operation $-3*X/2.4+5.5$.

```
>> -3*X/2.4 + 5.5

ans =

        4.2500      8.0000      5.5000      4.2500
        3.0000      1.7500     -2.0000      3.0000
        9.2500     -0.7500      3.0000      4.2500
       -0.7500      8.0000      0.5000      0.5000
```

20. Determine the sine, cosine, and tangent of the matrix B given below (element-by-element).

$$B = \begin{bmatrix} \dfrac{\pi}{3} & \dfrac{2\pi}{3} \\ \dfrac{2\pi}{3} & \pi \end{bmatrix}$$

```
>> B = [pi/3 2*pi/3 ; 2*pi/3 pi]

B =

        1.0472      2.0944
        2.0944      3.1416

>> sin(B)

ans =

        0.8660      0.8660
        0.8660      0.0000

>> cos(B)

ans =
```

```
     0.5000     -0.5000
    -0.5000     -1.0000

>> tan(B)

ans =

     1.7321     -1.7321
    -1.7321     -0.0000
```

21. In Exercise 20 above, determine the square root of the matrix B element-by-element.

```
>> sqrt(B)

ans =

     1.0233      1.4472
     1.4472      1.7725
```

22. In Exercise 20 above, determine the true square root of the matrix B.

```
>> sqrtm(B)

ans =

     0.5821 + 0.3598i     0.9419 - 0.2224i
     0.9419 - 0.2224i     1.5240 + 0.1374i
```

Notice that we obtained a complex matrix. Try to check this answer by multiplying this complex matrix by itself to get the original matrix.

23. In Exercise 20 above, determine the exponential of the matrix B element-by-element.

```
>> exp(B)
```

```
ans =

    2.8497      8.1205
    8.1205     23.1407
```

24. In Exercise 20 above, determine the true exponential of the matrix B.

```
>> expm(B)

ans =

   23.9028     37.4119
   37.4119     61.3147
```

25. In Exercise 20 above, determine the natural logarithm of the matrix B element-by-element.

```
>> log(B)

ans =

    0.0461      0.7393
    0.7393      1.1447
```

26. In Exercise 20 above, determine the true natural logarithm of the matrix B.

```
>> logm(B)
Warning: Principal matrix logarithm is not
defined for A with
          nonpositive   real   eigenvalues.   A
non-principal matrix
          logarithm is returned.
> In funm at 153
   In logm at 27

ans =

   -0.5995 + 2.2733i    1.2912 - 1.4050i
```

```
   1.2912 - 1.4050i    0.6917 + 0.8683i
```

Notice that we obtain a non-principal complex matrix.

27. In Exercise 20 above, perform the exponential operation 4^B.

```
>> 4^B

ans =

   130.0124   209.2159
   209.2159   339.2283
```

28. In Exercise 27 above, repeat the same exponential operation but this time element-by-element.

```
>> 4.^B

ans =

     4.2705    18.2369
    18.2369    77.8802
```

29. In Exercise 20 above, perform the operation B^4.

```
>> B^4

ans =

   107.0297   173.1717
   173.1717   280.2015
```

30. Generate a rectangular matrix of 1's of size 2x3.

```
>> ones(2,3)

ans =

     1      1      1
```

```
        1        1        1
```

31. Generate a rectangular matrix of 0's of size 2x3.

```
>> zeros(2,3)

ans =

        0        0        0
        0        0        0
```

32. Generate a rectangular identity matrix of size 2x3.

```
>> eye(2,3)

ans =

        1        0        0
        0        1        0
```

33. Generate a square matrix of 1's of size 4.

```
>> ones(4)

ans =

        1        1        1        1
        1        1        1        1
        1        1        1        1
        1        1        1        1
```

34. Generate a square matrix of 0's of size 4.

```
>> zeros(4)

ans =

        0        0        0        0
        0        0        0        0
        0        0        0        0
```

```
        0        0        0        0
```

35. Generate a square identity matrix of size 4.

```
>> eye(4)

ans =

    1        0        0        0
    0        1        0        0
    0        0        1        0
    0        0        0        1
```

36. Determine the transpose of the following matrix:

$$C = \begin{bmatrix} 1 & 2 & -3 & 0 \\ 2 & 5 & 2 & -3 \\ 1 & 3 & 7 & -2 \\ 2 & 3 & -1 & 3 \end{bmatrix}$$

```
>> C = [1 2 -3 0 ; 2 5 2 -3 ; 1 3 7 -2 ; 2
   3 -1 3]

C =

    1        2       -3        0
    2        5        2       -3
    1        3        7       -2
    2        3       -1        3

>> C'

ans =

    1        2        1        2
    2        5        3        3
   -3        2        7       -1
    0       -3       -2        3
```

37. In Exercise 36 above, perform the operation $C + C'$. Do you get a symmetric matrix?

```
>> C + C'

ans =

    2     4    -2     2
    4    10     5     0
   -2     5    14    -3
    2     0    -3     6
```

We get a symmetric matrix.

38. In Exercise 36 above, extract the diagonal of the matrix C.

```
>> diag(C)

ans =

    1
    5
    7
    3
```

39. In Exercise 36 above, extract the upper triangular part and the lower triangular part of the matrix C.

```
>> triu(C)

ans =

    1     2    -3     0
    0     5     2    -3
    0     0     7    -2
    0     0     0     3

>> tril(C)
```

ans =

1	0	0	0
2	5	0	0
1	3	7	0
2	3	-1	3

40. In Exercise 36 above, determine the determinant and trace of the matrix C. Do you get scalars?

```
>> det(C)
```

ans =

   -13

```
>> trace(C)
```

ans =

   16

Both the determinant and trace are scalars (numbers in this example).

41. In Exercise 36 above, determine the inverse of the matrix C.

```
>> inv(C)
```

ans =

-10.5385	6.3846	-6.0000	2.3846
5.3077	-3.0769	3.0000	-1.0769
-0.3077	0.0769	-0.0000	0.0769
1.6154	-1.1538	1.0000	-0.1538

42. In Exercise 41 above, multiply the matrix C by its inverse matrix. Do you get the identity matrix?

```
>> C*inv(C)
```

```
ans =

    1.0000      0.0000      0.0000      0.0000
   -0.0000      1.0000           0      0.0000
         0      0.0000      1.0000           0
    0.0000      0.0000           0      1.0000
```

We get the identity matrix.

43. In Exercise 36 above, determine the norm of the matrix C.

```
>> norm(C)
```

```
ans =

    9.5800
```

44. In Exercise 36 above, determine the eigenvalues of the matrix C.

```
>> eig(C)
```

```
ans =

   -0.0702
    3.9379 + 2.6615i
    3.9379 - 2.6615i
    8.1944
```

We get four eigenvalues – some of them are complex numbers.

45. In Exercise 36 above, determine the coefficients of the characteristic polynomial of the matrix C.

```
>> poly(C)
```

```
ans =

    1.0000   -16.0000    86.0000  -179.0000
  -13.0000
```

We get a fourth-degree polynomial with five coefficients.

46. In Exercise 36 above, determine the rank of the matrix C.

```
>> rank(C)

ans =

     4
```

47. Generate a square random matrix of size 5.

```
>> rand(5)

ans =

    0.9501    0.7621    0.6154    0.4057
0.0579
    0.2311    0.4565    0.7919    0.9355
0.3529
    0.6068    0.0185    0.9218    0.9169
0.8132
    0.4860    0.8214    0.7382    0.4103
0.0099
    0.8913    0.4447    0.1763    0.8936
0.1389
```

48. Generate a square magic matrix of size 7. What is the sum of each row, column or diagonal in this matrix.

```
>> magic(7)

ans =

    30    39    48     1    10    19    28
    38    47     7     9    18    27    29
    46     6     8    17    26    35    37
     5    14    16    25    34    36    45
    13    15    24    33    42    44     4
```

$$
\begin{array}{ccccccc}
21 & 23 & 32 & 41 & 43 & 3 & 12 \\
22 & 31 & 40 & 49 & 2 & 11 & 20
\end{array}
$$

The sum of each row, column or diagonal is 175.

49. Generate the following two symbolic matrices:

$$
X = \begin{bmatrix} 1 & x \\ x-2 & x^2 \end{bmatrix} \quad , \quad Y = \begin{bmatrix} \dfrac{x}{2} & 3x \\ 1-x & 4 \end{bmatrix}
$$

```
>> syms x
>> X = [1 x ; x-2 x^2]

X =

[    1,    x]
[ x-2,  x^2]


>> Y = [x/2 3*x ; 1-x 4]

Y =

[ 1/2*x,    3*x]
[    1-x,      4]
```

50. In Exercise 49 above, perform the matrix subtraction operation $X - Y$ to obtain the new matrix Z.

```
>> Z = X-Y

Z =

[ 1-1/2*x,      -2*x]
[    2*x-3,    x^2-4]
```

51. In Exercise 50 above, determine the transpose of the matrix Z.

```
>> Z'

ans =

[ 1-1/2*conj(x),   -3+2*conj(x)]
[    -2*conj(x),   -4+conj(x)^2]
```

In the above, it is assumed that x is a complex variable.

52. In Exercise 50 above, determine the trace of the matrix Z.

```
>> trace(Z)

ans =

-3-1/2*x+x^2
```

53. In Exercise 50 above, determine the determinant of the matrix Z.

```
>> det(Z)

ans =

5*x^2-4-1/2*x^3-4*x
```

54. In Exercise 50 above, determine the inverse of the matrix Z.

```
>> inv(Z)

ans =

[ -2*(x^2-4)/(-10*x^2+8+x^3+8*x),
-4*x/(-10*x^2+8+x^3+8*x)]
[   2*(2*x-3)/(-10*x^2+8+x^3+8*x),
(x-2)/(-10*x^2+8+x^3+8*x)]
```

The inverse of Z obtained above may be simplified by factoring out the determinant that appears common in the denominators of all the elements of the matrix.

8. Programming

1. Write a script of four lines as follows: the first line should be a comment line, the second and third lines should have the assignments `cost = 200` and `sale_price = 250`, respectively. The fourth line should have the calculation `profit = sale_price - cost`. Store the script in a script file called `example8.m`. Finally run the script file.

```
% This is an example
cost = 200
sale_price = 250
profit = sale_price - cost
```

Now, run the above script as follows:

```
>> example8

cost =

    200


sale_price =

    250


profit =

    50
```

2. Write a function of three lines to calculate the volume of a sphere of radius `r`. The first line should include the name of the function which is `volume(r)`. The second line should be a comment line. The third line should include the calculation of the volume of the sphere which is $\frac{4}{3}\pi r^3$. Store the function in a function file called

volume.m then run the function with the value of r equal to 2 (no units are used in this exercise).

```
function volume(r)
% This is a function
volume = (4/3)*pi*r^3
```

Now, run the above function as follows:

```
>> volume(2)

volume =

   33.5103
```

3. Write a function with two arguments to calculate the area of a rectangle with sides a and b. The function should have three lines. The first line should include the name of the function which is RectangleArea(a,b). The second line should be a comment line. The third line should include the calculation of the area of the rectangle with is the product a*b. Store the function in a function file called RectangleArea.m then run the function twice as follow: the first execution with the values 3 and 6, while the second execution with the values 2.5 and 5.5.

```
function RectangleArea(a,b)
% This is another example of a function
RectangleArea = a*b
```

Now, run the above function as follows:

```
>> RectangleArea(3,6)

RectangleArea =

   18

>> RectangleArea(2.5,5.5)

RectangleArea =
```

13.7500

4. Write a script containing a For loop to compute the vector x to have the values $x(n) = n^3$ where n has the range from 1 to 7. Include a comment line at the beginning. Store the script in a script file called example9.m then run the script and display the values of the elements of the vector x.

```
% This is an example of a FOR loop
for n = 1:7
    x(n) = n^3;
end
```

Now, run the above script as follows:

```
>> example9
>> x

x =

    1     8    27    64   125   216   343
```

5. Write a script containing two nested For loops to compute the matrix y to have the values $y(m,n) = m^2 - n^2$ where both m and n each has the range from 1 to 4. Include a comment line at the beginning. Store the script in a script file called example10.m then run the script and display the values of the elements of the matrix y.

```
%This is an example of another FOR loop
for n = 1:4
    for m = 1:4
        y(m,n) = m^2 - n^2;
    end
end
```

Now, run the above script as follows:

```
>> example10
>> y

y =

        0      -3      -8     -15
        3       0      -5     -12
        8       5       0      -7
       15      12       7       0
```

6. Write a script containing a While loop using the two variables tol and n. Before entering the While loop, initialize the two variables using the assignments tol = 0.0 and n = 3. Then use the two computations n = n + 1 and tol = tol + 0.1 inside the While loop. Make the loop end when the value of tol becomes equal or larger than 1.5. Include a comment line at the beginning. Store the script in a script file called example11.m then run the script and display the values of the two variables tol and n.

```
% This is an example of a While loop
tol = 0.0;
n = 3;
while tol < 1.5
    n = n + 1;
    tol = tol + 0.1;
end
```

Now, run the above script as follows:

```
>> example11
>> n

n =

    18

>> tol

tol =
```

1.5000

7. Write a function called price(items) containing an If construct as follows. Let the price of the items be determined by the computation price = items*130 unless the value of the variable items is greater than 5 – then in this case the computation price = items*160 should be used instead. Include a comment line at the beginning. Store the function in a function file called price.m then run the function twice with the values of 3 and 9 for the variable items. Make sure that the function displays the results for the variable price.

```
function price(items)
% This is an example of an If Elseif construct
price = items*130
if items > 5
    price = items*160
end
```

Now, run the above function as follows:

```
>> price(3)

price =

    390

>> price(9)

price =

        1170


price =

        1440
```

8. Write a function called price2(items) containing an If Elseif construct as follows. If the value of the variable items is less than 3, then compute the variable price2 by multiplying items by 130. In the second case, if the value of the variable items is less than 5, then compute the variable price2 by multiplying items by 160. In the last case, if the value of the variable items is larger than 5, then compute the variable price2 by multiplying the items by 200. Include a comment line at the beginning. Store the function in a function file called price2.m then run the function three times – with the values of 2, 4, and 6. Make sure that the function displays the results for the variable price2.

```
function price2(items)
% This is an example of an If Elseif construct
if items < 3
    price2 = items*130
elseif items < 5
    price2 = items*160
elseif items > 5
    price2 = items*200
end
```

Now, run the above function as follows:

```
>> price2(2)

price2 =

   260

>> price2(4)

price2 =

   640

>> price2(6)

price2 =
```

```
        1200
```

9. Write a function called price3(items) containing a Switch
    Case construct. The function should produce the same results
    obtained in Exercise 8 above. Include a comment line at the
    beginning. Store the function in a function file called price3.m
    then run the function three times – with the values of 2, 4, and 6.
    Make sure that the function displays the results for the variable
    price3.

```
        function price3(items)
        % This is an example of the Switch Case construct
        switch items
            case 2
                price3 = items* 130
            case 4
                price3 = items* 160
            case 6
                price3 = items* 200
            otherwise
                price3 = 0
        end
```

Now, run the above function as follows:

```
>> price3(2)

price3 =

    260

>> price3(4)

price3 =

    640

>> price3(6)
```

```
price3 =

        1200
```

10. Write a script file to store the following symbolic matrix A then calculate its third power $B = A^3$. Include a comment line at the beginning. Store the script in a script file called example12.m then run the script to display the two matrices A and B.

$$A = \begin{bmatrix} \dfrac{x}{2} & 1-x \\ x & 3x \end{bmatrix}$$

```
% This is an example with the Symbolic
        Math Toolbox
syms x
A = [x/2 1-x ; x 3*x]
B = A^3
```

Now, run the above script as follows:

```
>> example12

A =

[ 1/2*x,    1-x]
[     x,    3*x]



B =

[      1/2*x*(1/4*x^2+(1-x)*x)+7/2*(1-x)*x^2,
       7/4*(1-x)*x^2+(1-x)*((1-x)*x+9*x^2)]
[          x*(1/4*x^2+(1-x)*x)+21/2*x^3,
       7/2*(1-x)*x^2+3*x*((1-x)*x+9*x^2)]
```

11. Write a function called SquareRoot2 (matrix) similar to the function SquareRoot (matrix) described at the end of this

chapter but with the following change. Substitute the value of 1.5 instead of 1 for the symbolic variable x. Make sure that you include a comment line at the beginning. Store the function in a function file called SquareRoot2.m then run the function using the following symbolic matrix:

$$M = \begin{bmatrix} 2 & x & 0 \\ 3-x & 5 & -x \\ x+2 & 1 & 3 \end{bmatrix}$$

```
function SquareRoot2(matrix)
% This is an example of a function
        with the Symbolic Math Toolbox
y = det(matrix)
z = subs(y,1.5)
if z < 1
    M = 2*sqrt(matrix)
else
    M = sqrt(matrix)
end
```

Now, run the above function as follows:

```
>> syms x
>> M = [2 x 0 ; 3-x 5 -x ; x+2 1 3]

M =

[   2,    x,    0]
[ 3-x,    5,   -x]
[ x+2,    1,    3]


>> SquareRoot2(M)

y =

30-7*x+x^2-x^3
```

z =

   18.3750

M =

   [2^(1/2),   x^(1/2),   0]
   [(3-x)^(1/2),   5^(1/2),   (-x)^(1/2)]
   [(x+2)^(1/2),   1,   3^(1/2)]

9. Graphs

Solve all the exercises using MATLAB. All the needed MATLAB commands for these exercises were presented in this chapter.

1. Plot a two-dimensional graph of the two vectors x = [1 2 3 4 5 6 7] and y = [10 15 23 43 30 10 12] using the `plot` command.

    ```
    >> x = [1 2 3 4 5 6 7]
    ```

    x =

            1   2   3   4   5   6   7

    ```
    >> y = [10 15 23 43 30 10 12]
    ```

    y =

           10   15   23   43   30   10   12

    ```
    >> plot(x,y)
    ```

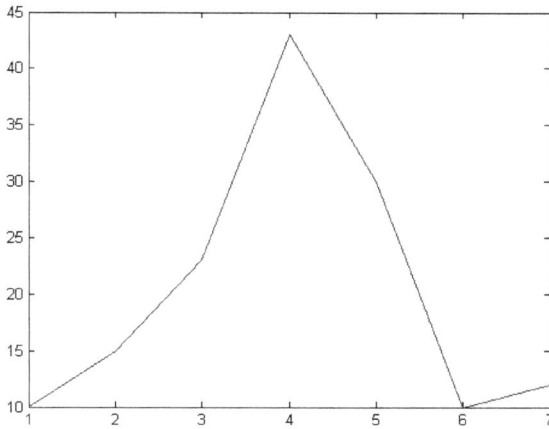

2.   In Exercise 1 above, add to the graph a suitable title along with labels for the x-axis and the y-axis.

```
>> hold on;
>> title('This is an exercise')
>> xlabel('x-axis')
>> ylabel('y-axis')
```

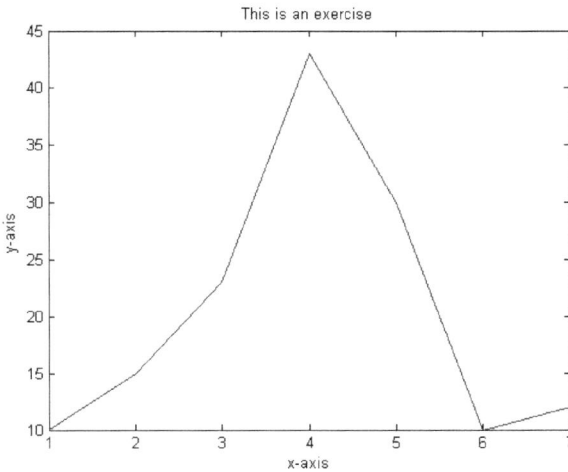

3. Plot the mathematical function $y = 2x^3 + 5$ in the range x from
    -6 to $+6$. Include a title for the graph as well as labels for the two
    axes.

```
>> x = [-6 -5 -4 -3 -2 -1 0 1 2 3 4 5 6];
>> y = 2*x.^3 +5;
>> plot(x,y)
>> hold on;
>> title('This is another exercise')
>> xlabel('values of x')
>> ylabel('values of y')
```

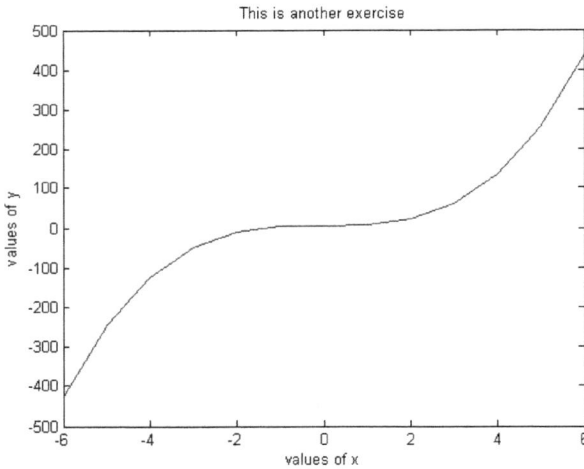

4. Repeat the plot of Exercise 3 above but show the curve with a blue
    dashed line with circle symbols at the plotted points.

```
plot(x,y,'b--o')
```

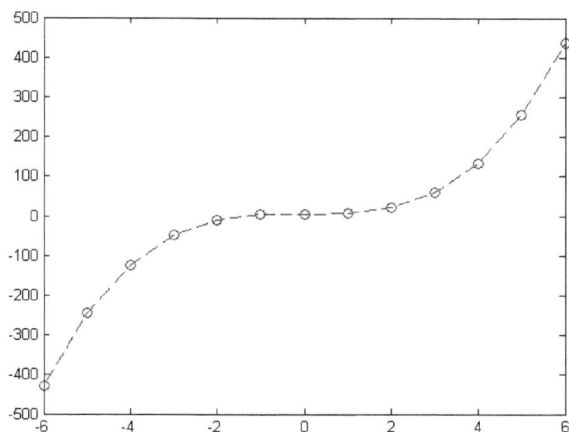

5. Plot the two mathematical functions $y = 2\sin(\dfrac{x}{3})$ and $z = 2\cos(\dfrac{x}{3})$ such that the two curves appear on the same diagram. Make the range for x between 0 and 3π with increments of $\dfrac{\pi}{4}$. Distinguish the two curves by plotting one with a dashed line and the other one with a dotted line. Include the title and axis information on the graph as well as a grid and a legend.

```
>> x = 0:pi/4:3*pi;
>> y = 2*sin(x/3);
>> z = 2*cos(x/3);
>> plot(x,y,'--',x,z,':')
>> hold on;
>> grid on;
>> title('This is an exercise in plotting two curves
on the same diagram')
>> xlabel('x')
>> ylabel('2sin(x/3) and 2cos(x/3)')
>> legend('2sin(x/3)','2cos(x/3)')
```

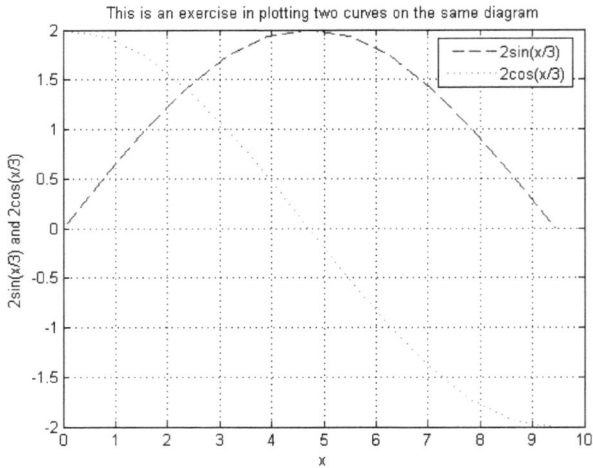

This is an exercise in plotting two curves on the same diagram

6. Plot the following four mathematical functions each with its own diagram using the subplot command. The functions are $y = 2x^3 - 4$, $z = x + 1$, $w = 2 - \sqrt{x}$, and $v = x^2 + 3$. Use the vector x = [1 2 3 4 5 6 7 8 9 10] as the range for x. No need to show title or axis information.

```
>> x = [1 2 3 4 5 6 7 8 9 10];
>> y = 2*x.^3 -4;
>> z = x+1;
>> w = 2 - sqrt(x);
>> v = x.^2 +3;
>> subplot(2,2,1);
>> hold on;
>> plot(x,y);
>> subplot(2,2,2);
>> plot(x,z);
>> subplot(2,2,3);
>> plot(x,w);
>> subplot(2,2,4);
>> plot(x,v);
```

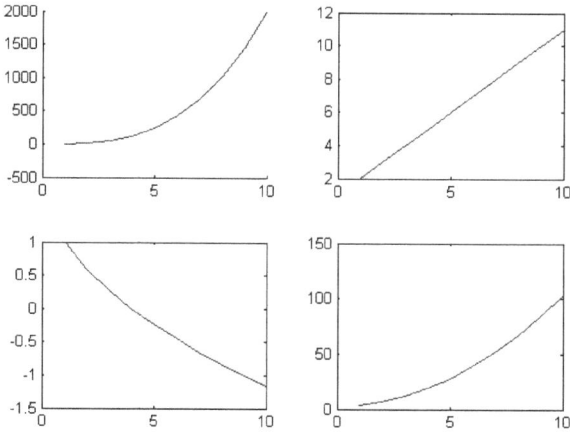

7. Use the plot3 command to show a three-dimensional curve of the equation $z = 2\sin(xy)$. Use the vector x = [1 2 3 4 5 6 7 8 9 10] for both x and y. No need to show title or axis information.

```
>> x = [1 2 3 4 5 6 7 8 9 10];
>> y = [1 2 3 4 5 6 7 8 9 10];
>> z = 2*sin(x.*y);
>> plot3(x,y,z)
```

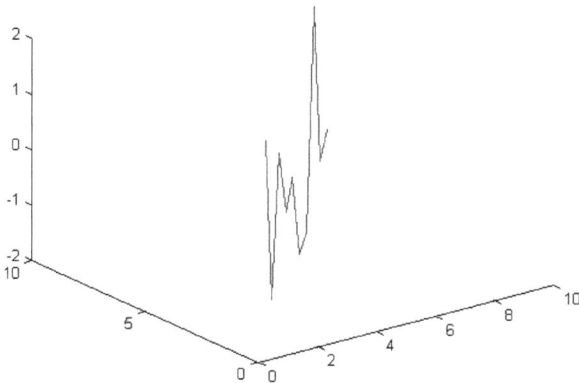

8.   Use the mesh command to plot a three-dimensional mesh surface of elements of the matrix M given below. No need to show title or axis information.

$$M = \begin{bmatrix} 0.1 & 0.2 & 0.5 & 0.7 & 0.3 \\ 0.2 & 0.8 & 0.5 & 0.6 & 0.3 \\ 0.5 & 0.9 & 0.9 & 0.4 & 0.4 \\ 0.4 & 0.4 & 0.5 & 0.7 & 0.9 \\ 0.1 & 0.3 & 0.4 & 0.6 & 0.8 \end{bmatrix}$$

```
>> M = [0.1 0.2 0.5 0.7 0.3 ; 0.2 0.8 0.5 0.6 0.3
; 0.5 0.9 0.9 0.4 0.4 ; 0.4 0.4 0.5 0.7 0.9 ; 0.1
0.3 0.4 0.6 0.8]

M =

      0.1000        0.2000        0.5000        0.7000
0.3000
      0.2000        0.8000        0.5000        0.6000
0.3000
      0.5000        0.9000        0.9000        0.4000
0.4000
      0.4000        0.4000        0.5000        0.7000
0.9000
      0.1000        0.3000        0.4000        0.6000
0.8000

>> mesh(M)
```

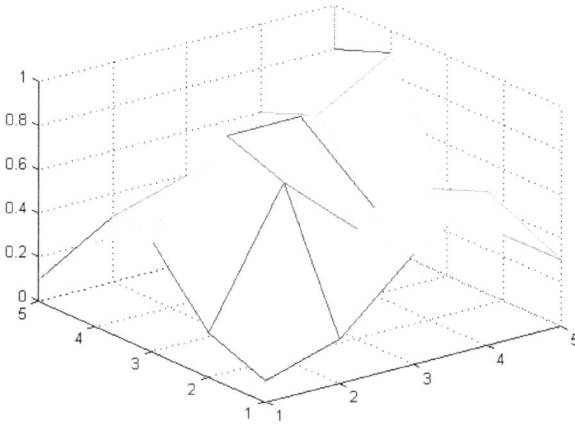

9. Use the `surf` command to plot a three-dimensional surface of the elements of the matrix M given in Exercise 8 above. No need to show title or axis information.

```
>> surf(M)
```

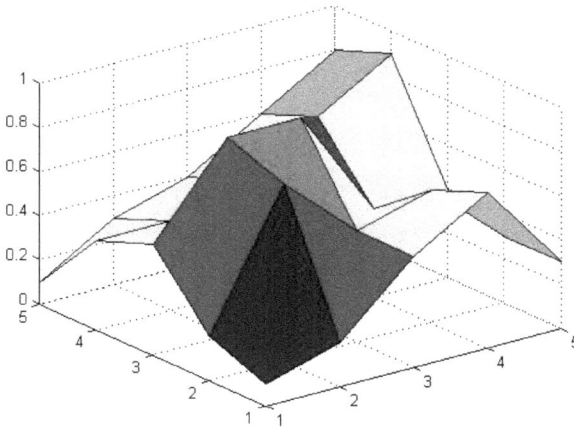

10. Use the `contour` command to plot a contour map of the elements of the matrix M given in Exercise 8 above. No need to show the contour values on the graph.

```
>> contour(M)
```

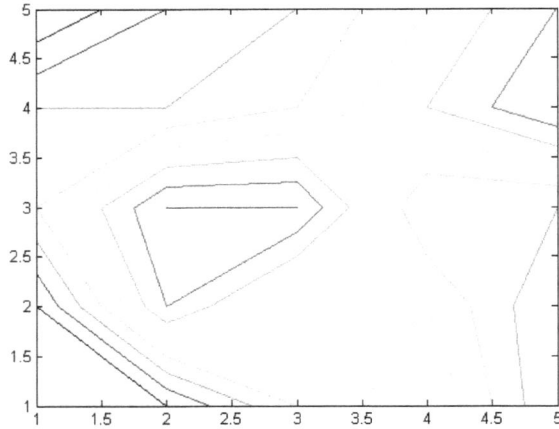

11. Use the `surfc` command to plot a three-dimensional surface along with contours underneath it of the matrix M given in Exercise 8 above. No need to show title or axis information.

```
>> surfc(M)
```

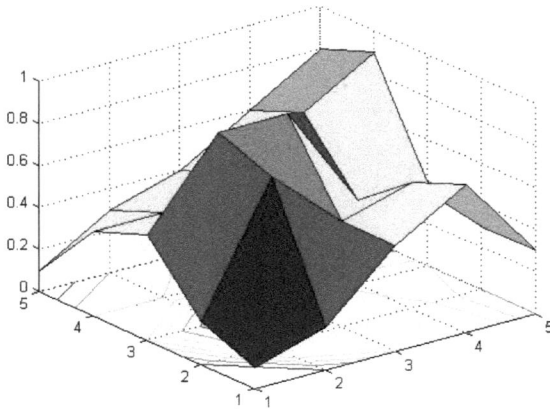

10. Solving Equations

1. Solve the following linear algebraic equation for the variable x. Use the roots command.

    $$3x + 5 = 0$$

    ```
    >> p = [3 5]

    p =

         3      5

    >> roots(p)

    ans =

       -1.6667
    ```

2. Solve the following quadratic algebraic equation for the variable x. Use the roots command.

    $$x^2 + x + 1 = 0$$

    ```
    >> p = [1 1 1]

    p =

         1      1      1

    >> roots(p)

    ans =

       -0.5000 + 0.8660i
       -0.5000 - 0.8660i
    ```

3. Solve the following algebraic equation for x. Use the roots command.

$$3x^4 - 2x^3 + x - 3 = 0$$

```
>> p = [3 -2 0 1 -3]

p =

    3    -2     0     1    -3

>> roots(p)

ans =

    1.1170
    0.2426 + 0.9477i
    0.2426 - 0.9477i
   -0.9355
```

4. Solve the following system of linear simultaneous algebraic equations for the variables x and y. Use the inverse matrix method.

$$3x + 5y = 9$$
$$4x - 7y = 13$$

```
>> A = [3 5 ; 4 -7]

A =

    3     5
    4    -7

>> b = [9 ; 13]

b =

    9
   13
```

```
>> x = inv(A)*b

x =

    3.1220
   -0.0732
```

5. In Exercise 4 above, solve the same linear system again but using the Gaussian elimination method with the backslash operator.

```
>> A = [3 5 ; 4 -7]

A =

    3       5
    4      -7

>> b = [9 ; 13]

b =

    9
   13

>> x = A\b

x =

    3.1220
   -0.0732
```

6. Solve the following system of linear simultaneous algebraic equations for the variables x, y, and z. Use Gaussian elimination with the backslash operator.

$$2x - y + 3z = 5$$
$$4x + 5z = 12$$
$$x + y + 2z = -3$$

```
>> A = [2 -1 3 ; 4 0 5 ; 1 1 2]

A =

        2      -1       3
        4       0       5
        1       1       2

>> b = [5 ; 12 ; -3]

b =

        5
       12
       -3

>> x = A\b

x =

    10.0000
    -1.8000
    -5.6000
```

7. Solve the following linear algebraic equation for the variable x in terms of the constant a.

$$2x + a = 5$$

```
>> syms x
>> syms a
>> solve('2*x + a - 5 = 0')

ans =

-1/2*a+5/2
```

8. Solve the following quadratic algebraic equation for the variable x in terms of the constants a and b.

$$x^2 + ax + b = 0$$

```
>> syms x
>> syms a
>> syms b
>> solve('x^2 + a*x + b = 0')

ans =

  -1/2*a+1/2*(a^2-4*b)^(1/2)
  -1/2*a-1/2*(a^2-4*b)^(1/2)
```

9. Solve the following nonlinear equation for the variable x.

$$2e^x + 3\cos x = 0$$

```
>> syms x
>> solve('2*exp(x) + 3*cos(x) = 0')

ans =

-1.6936679094546100269188394124169
```

The above answer is approximated as -1.694.

10. Solve the following system of linear simultaneous algebraic equations for the variables x and y in terms of the constant c.

$$2x - 3cy = 5$$
$$cx + 2y = 7$$

```
>> syms x
>> syms y
>> syms c
>> [x,y]  =  solve('2*x  -  3*c*y  -  5 = 0',
        'c*x + 2*y - 7 = 0')

x =
```

```
(21*c+10)/(4+3*c^2)

y =

-1/(4+3*c^2)*(-14+5*c)
```

11. Solve the following system of nonlinear simultaneous algebraic equations for the variables x and y.

$$3x^2 - 2x + y = 7$$
$$xy + x = 5$$

```
>> syms x
>> syms y
>> [x,y] = solve('3*x^2 - 2*x + y - 7 = 0',
     'x*y + x - 5 = 0')

x =

            5/3
   1/2*5^(1/2)-1/2
  -1/2-1/2*5^(1/2)

y =

            2
  3/2+5/2*5^(1/2)
  3/2-5/2*5^(1/2)
```

11. Beginning Calculus

1. Define the following mathematical function in MATLAB using the inline command:

$$f(x) = 3x^2 + x - 1$$

```
>> f = inline('3*x^2 + x - 1', 'x')
```

```
f =

    Inline function:
    f(x) = 3*x^2 + x - 1
```

2. In Exercise 1 above, evaluate the function f at $x = 1$.

```
>> f(1)

ans =

    3
```

3. In Exercise 1 above, evaluate the function f at $x = -2$.

```
>> f(-2)

ans =

    9
```

4. In Exercise 1 above, differentiate the function f with respect to x.

```
>> syms x
>> diff(f(x),x)

ans =

6*x+1
```

5. Define the following mathematical function in MATLAB using the inline command:

$$g(y) = 2\sin(\pi y) + 3y\cos(\pi y)$$

```
>>      g      =      inline('2*sin(pi*y)      +
3*y*cos(pi*y)',  'y')
```

```
g =

    Inline function:
    g(y) = 2*sin(pi*y) + 3*y*cos(pi*y)
```

6. In Exercise 5 above, differentiate the function g with respect to y.

```
>> diff(g(y),y)

ans =

2*cos(pi*y)*pi+3*cos(pi*y)-3*y*sin(pi*y)*pi
```

7. In Exercise 5 above, find the indefinite integral of the function g.

```
>> int(g(y))

ans =

-
2/pi*cos(pi*y)+3/pi^2*(cos(pi*y)+y*sin(pi*y
)*pi)
```

8. In Exercise 5 above, find the value of the following definite integral

$$\int_0^1 g(y)dy$$

```
>> int(g(y),0,1)

ans =

2*(2*pi-3)/pi^2
```

9. In Exercise 8 above, evaluate the value obtained numerically using the `double` command.

```
>> double(ans)

ans =
```

```
      0.6653
```

10. Evaluate the following limit in MATLAB:

$$\lim_{x \to 0}(\sin x + \cos x)$$

```
>> limit(sin(x)+cos(x),x,0)

ans =

1
```

11. Evaluate the following limit in MATLAB:

$$\lim_{x \to \infty}\frac{x^2 + x + 1}{3x^2 - 2}$$

```
>> limit((x^2+x+1)/(3*x^2-2),x,Inf)

ans =

1/3
```

12. Find the Taylor series expansion for the function $\cos x$ up to eight terms.

```
>> taylor(cos(x),x,8)

ans =

1-1/2*x^2+1/24*x^4-1/720*x^6
```

13. Find the Taylor series expansion for the function e^x up to nine terms.

```
>> taylor(exp(x),x,9)
```

```
ans =

1+x+1/2*x^2+1/6*x^3+1/24*x^4+1/120*x^5+1/72
0*x^6+1/5040*x^7+1/40320*x^8
```

14. Evaluate the following sum symbolically using the symsum command:

$$\sum_{1}^{n} \frac{1}{k}$$

```
>> syms k
>> syms n
>> symsum(1/k,1,n)

ans =

Psi(n+1)+eulergamma
```

The above result is written in terms of special functions. Consult a book on special function for more details.

15. Solve the following initial value ordinary differential equation using the dsolve command:

$$\frac{dy}{dx} = xy - \sin x + 3 \qquad , \qquad y(0) = 0$$

```
>>    dsolve('Dy    =    x*y    -    sin(x)    +    3',
'y(0)=0')

ans =

(sin(x)-3+exp(x*t)*(-sin(x)+3))/x
```

References

MATLAB Tutorials

1. http://www.cyclismo.org/tutorial/matlab/

   This is an online tutorial in HTML format from Union College, New York. The tutorial emphasizes vectors, matrices, vector operations, loops, plots, executable files (scripts), subroutines (functions), if statements, and data files.

2. http://www.math.ufl.edu/help/matlab-tutorial/

   This is an online tutorial in HTML format from the University of Florida, Gainesville. The tutorial emphasizes matrices, variables, functions, decisions, loops, and scripts. This tutorial has a nice list of important commands of MATLAB with a short description of each command.

3. http://www.mines.utah.edu/gg_computer_seminar/matlab/matlab.html

   This is an online tutorial in HTML format available at the website of University of Utah, Salt Lake City (Original by Kermit Sigmon, University of Florida). This tutorial emphasizes matrices, decisions, loops, scalar functions, vector functions, matrix functions, strings, and graphics. You may also download the tutorial as a postscript file (39 pages that include a comprehensive reference section).

4. http://web.mit.edu/afs/.athena/astaff/project/logos/olh/Math/Matlab/Matlab.html

This is an online tutorial in HTML format hosted at MIT. It emphasizes matrices, arithmetic and logical operators, control structures, selective indexing, polynomial operations, signal processing functions, graphics, scripts, and functions.

5. http://www.mit.edu/people/abbe/matlab/main.html

This is an online 3-day tutorial at MIT in HTML format. It covers matrices, various MATLAB commands, help, plotting, polynomials, variables, scripts, functions, loops, debugging, differential equations, vectorization, three-dimensional graphics, and symbolic math.

6. http://texas.math.ttu.edu/~gilliam/ttu/m4330/m4330.html

This is a tutorial hosted at Texas Tech University, Lubbock. They provide six PDF files for download in the form of six lessons for a mathematics course they teach there. These lessons are somewhat advanced and specialized and are not recommended for beginner students of MATLAB.

7. http://www.engin.umich.edu/group/ctm/basic/basic.html

This is a basic tutorial from Carnegie Mellon and the University of Michigan. It covers the basics of vectors, functions, plotting, polynomials, matrices, printing, help, and M-files.

8. http://www.mathworks.com/academia/student_center/tutorials/launchpad.html

This is an online tutorial from the MathWorks, the company that develops and sells MATLAB. This tutorial covers variables, calculations, plotting, scripts, and files.

9. http://www.math.utah.edu/lab/ms/matlab/matlab.html#starting

This is an online short tutorial from the University of Utah, Salt Lake City. This is a very simple tutorial recommended for beginners. It covers matrices, vectors, systems of equations, loops, and graphing in one and two dimensions.

10. http://www.math.mtu.edu/~msgocken/intro/intro.html

This is an online tutorial in HTML format hosted at Michigan Tech and written by Mark S. Gockenback. It covers calculations, graphs, programming, advanced matrix calculations, advanced graphics, solving nonlinear problems, and advanced data types like structures, cell arrays, and objects.

11. http://www.eece.maine.edu/mm/matweb.html

MATLAB Educational Websites: This is an online HTML page containing links to several educational websites on various topics using MATLAB. The page contains also links to numerous MATLAB tutorials.

12. http://web.cecs.pdx.edu/~gerry/MATLAB/

MATLAB Hypertext Reference: This is an online tutorial in HTML format written by Gerald Recktenwald and hosted at Portland State University, Portland. This tutorial covers variables, plotting, and programming.

13. http://www.facstaff.bucknell.edu/maneval/help211/helpmain.html

Helpful Information for Using MATLAB: These pages of helpful information are maintained by Jim Maneval and hosted at Bucknell University, Lewisburg.

14. http://www.indiana.edu/~statmath/math/matlab/gettingsta
    rted/index.html

Getting Started with MATLAB: This is an online tutorial in
HTML format. The tutorial may also be downloaded and
printed as a PDF file (14 pages). This tutorial covers syntax,
matrices, graphics, and programming.

15. www.math.udel.edu/~driscoll/teaching/**matlab_adv**.pdf

Crash Course in MATLAB: This tutorial is in the form of a
PDF file download. It is written by Tobin A. Driscoll and
hosted at the University of Delaware, The tutorial is 66 pages
that can be downloaded and printed. This tutorial covers
arrays, matrices, scripts, functions, errors, 2D and 3D
graphics, color, handles and properties, vectorization,
advanced data structures (strings, cell arrays, structures),
linear algebra, optimization, data fitting, quadrature, and
differential equations.

16. www.maths.dundee.ac.uk/~ftp/na-
    reports/**MatlabNotes**.pdf

An Introduction to MATLAB: This tutorial is in the form of
a PDF file download. It is written by David F. Griffiths and is
hosted at the University of Dundee. The tutorial is 37 pages
that can be downloaded and printed. This tutorial covers
numbers, formats, variables, output, vectors, 2D plotting,
scripts, vector operations, matrices, matrix operations,
systems of linear equations, strings, loops, decisions, 3D
plotting, files, and graphical user interfaces.

17. www.geosci.uchicago.edu/~gidon/geosci236/organize/**matl
    ab**Intro.pdf

An Introduction to MATLAB: This tutorial is in the form of
a PDF file download. It is written by S. Butenko, P. Pardalos,

and L. Pitsoulis. It is hosted at the University of Chicago. The tutorial is 28 pages that can be downloaded and printed. This tutorial covers matrices, matrix operations, workspace, formats, functions, vector functions, matrix functions, polynomial functions, programming, loops, decisions, M-files, and 2D and 3D graphics.

18. http://www.math.colostate.edu/~gerhard/classes/340/notes/index.html

MATLAB Notes: This tutorial is in the form HTML pages online. It is hosted at Colorado State University. This tutorial covers arithmetic, array operations, scripts, functions, plotting, polynomials, matrices, vectors, linear systems of algebraic equations, and symbolic computation. There are some links to more tutorials on the internet that are hosted at other places.

19. http://www.indiana.edu/~statmath/math/matlab/gettingstarted/index.html

Getting Started with MATLAB: This tutorial is in the form of online HTML pages that are hosted at Indiana University. It may also be downloaded as a PDF file (14 pages) and printed. This tutorial covers syntax, matrices, graphics, and programming.

20. http://www-ccs.ucsd.edu/matlab/toolbox/symbolic/symbmath.html

This is an online reference guide for the MATLAB Symbolic Math Toolbox. It is hosted at the University of California, San Diego. The tutorial covers about 85 MATLAB symbolic math commands that are listed alphabetically with examples.

21. http://www.phys.ufl.edu/docs/matlab/toolbox/symbolic/ch1.html

Using the Symbolic Math Toolbox: This tutorial is in the form of HTML pages that is hosted at the University of Florida, Gainesville. This tutorial seems to be comprehensive and covers calculus, simplifications, substitutions, variable-precision arithmetic, linear algebra, solving equations, special mathematical functions, using Maple functions, and the extended symbolic math toolbox.

22. http://en.wikipedia.org/wiki/Matlab

This is a small overview in the form of an HTML page hosted at the website of Wikipedia, the free internet encyclopedia. The page has a brief history of MATLAB and some useful links.

23. http://www.facstaff.bucknell.edu/maneval/help211/exercises.html

MATLAB Exercises: This web page provides some MATLAB exercises for practice. The site is hosted at Bucknell University. The exercises cover syntax, arrays, relational operators, logical operators, decisions, loops, and programming.

24. http://www.math.chalmers.se/~nilss/enm/INTRO_matlab_I.html

MATLAB Exercises: This web page provides three sets of MATLAB exercises for practice. The site is hosted at Chalmers University of Technology and Gteborg University. The exercises cover arithmetic, formats, variables, vectors, matrices, decisions, scripts, loops, computer arithmetic, strings, and graphics.

25. http://college.hmco.com/mathematics/larson/elementary_linear/5e/students/matlabs.html

MATLAB Exercises: These are MATLAB exercises taken from the book "Elementary Linear Algebra", fifth edition, by Ron Larson. Each set of exercises is downloaded as a PDF file that can be printed. These exercises cover systems of linear equations, matrices, determinants, vector spaces, inner product spaces, linear transformations, eigenvalues and eigenvectors, and complex vector spaces.

26. http://www.math.umn.edu/~ROberts/math5385/matlabEx 1.html

Supplementary MATLAB Exercises: These exercises are hosted at the University of Minnesota. They are written specifically for a mathematics course at the university. These exercises mostly deal with the graphics capabilities of MATLAB.

27. www.kom.aau.dk/~borre/**matlab**7/exercise.pdf

This is a PDF file download of a set of 8 pages of MATLAB exercises. The exercises are somewhat advanced and are not recommended for beginning students of MATLAB.

28. www.chaos.swarthmore.edu/courses/Phys50L_2006/**Matla b/MatlabExercises**.pdf

This is a PDF file download of 27 pages of MATLAB exercises. These exercises are written for a physics course and are somewhat advanced. They are not recommended for beginners.

MATLAB Books:

29. Gilat, A., *MATLAB: An Introduction with Applications*, Second Edition, John Wiley & Sons, 2004.

30. Pratap, R., *Getting Started with MATLAB 7: An Introduction for Scientists and Engineers*, Oxford University Press, 2005.

31. Hanselmann, D. and Littlefield, B., *Mastering MATLAB 7*, Prentice Hall, 2004.

32. Palm, W., *Introduction to MATLAB 7 for Engineers*, McGraw-Hill, 2004.

33. Moore, H., *MATLAB for Engineers*, Prentice Hall, 2006.

34. Chapman, S., *MATLAB Programming for Engineers*, Thomson Engineering, 2004.

35. Davis, T. and Sigmon, K., *MATLAB Primer*, Seventh Edition, Chapman & Hall, 2004.

36. Higham, D. and Higham, N., *MATLAB Guide*, Second Edition, SIAM, 2005.

37. King, J., *MATLAB for Engineers*, Addison-Wesley, 1988.

38. Etter, D., *Introduction to MATLAB for Engineers and Scientists*, Prentice Hall, 1995.

39. Magrab, E. et al., *An Engineer's Guide to MATLAB*, Prentice Hall, 2000.

40. Etter, D., Kuncicky, D. and Hull, D., *Introduction to MATLAB 6*, Prentice Hall, 2001.

41. Recktenwald, G., *Introduction to Numerical Methods and MATLAB: Implementation and Applications*, Prentice Hall, 2001.

42. Biran, A. and Breiner, M., *MATLAB 5 for Engineers*, Addison-Wesley, 1999.

43. Part-Enander, E. and Sjoberg, A., *The MATLAB 5 Handbook*, Addison-Wesley, 1999.

44. Etter, D., *Engineering Problem Solving with MATLAB*, Prentice Hall, 1993.

45. Chen, K., Giblin, P, and Irving, A., *Mathematical Explorations with MATLAB*, Cambridge University Press, 1999.

46. Mathews, J. and Fink, K., *Numerical Methods Using MATLAB*, Third Edition, Prentice Hall, 1999.

47. Fausett, L., *Applied Numerical Analysis Using MATLAB*, Prentice Hall, 1999.

Installation of MATLAB

In this book, it is assumed that you have already installed MATLAB on your computer system. To go through the eleven chapters of the book, you need to have MATLAB running on your computer. For help on installing MATLAB on your computer, check the following web links:

MATLAB Installation for Windows:

```
http://www.itc.virginia.edu/research/matlab/
     install/installwin.html
```

MATLAB Installation for Linux:

```
http://www.freebsd.org/doc/zh_TW/books/handb
     ook/linuxemu-matlab.html
```

MATLAB Installation for Macintosh:

```
http://www.itc.virginia.edu/research/matlab/
     install/installmac.html
```

MATLAB Installation for Other Systems:

```
http://shum.huji.ac.il/cc/matlR13linuxhomein
     st.html
```

Index

www.ingramcontent.com/pod-product-compliance
Ingram Content Group UK Ltd.
Pitfield, Milton Keynes, MK11 3LW, UK
UKHW020710220925
8009UKWH00042B/607

9 798869 055910